Clinical Experiences with Budipine in Parkinson Therapy

Edited by
F. Gerstenbrand, W. Poewe and G. Stern

With 108 Figures

Springer-Verlag
Berlin Heidelberg New York Tokyo

Professor Dr. Franz Gerstenbrand
Dr. Werner Poewe

Neurologische Universitätsklinik
Anichstraße 35, A-6020 Innsbruck

Dr. Gerald Stern

Department of Neurology
University College Hospital
Gower Street, London WC1E 6AU, Great Britain

ISBN-13: 978-3-540-13764-1 e-ISBN-13: 978-3-642-95455-9
DOI: 10.1007/978-3-642-95455-9

Library of Congress Cataloging in Publication Data. Clinical experiences with budipine in Parkinson therapy. Includes bibliographies and index. 1. Parkinsonism–Chemotherapy–Evaluation. 2. Budipine–Testing. 3. Antiparkinsonian agents. I. Gerstenbrand, F. (Franz) II. Poewe, W. (Werner) III. Stern, G. (Gerald), 1930-. [DNLM: 1. Antiparkinson Agents–therapeutic use. 2. Parkinson Disease–drug therapy. 3. Piperidines–therapeutic use. WL 359 C6402] RC382.C575 1985 616.8'33061 84-23626 ISBN 0-387-13764-5 (U.S.)

Typesetting, printing and bookbinding: Beltz Offsetdruck, Hemsbach
2125/3130-543210

Preface

Some 20 years ago the introduction of levodopa opened a new era in the treatment of Parkinson's disease. Although levodopa remains the single most effective drug for the treatment of this maladie, challenging therapeutic problems arise in many patients after several years of sustained treatment. Among these, fluctuations in motor performance and drug-induced involuntary movements are the most troublesome. The search for improved antiparkinsonian medications therefore continues and important new developments have already occurred, for example the introduction of the dopaminergic ergots and the selective MAO-B-inhibitor selegeline (deprenyl) into routine treatment.

The International Workshop on Parkinson's Disease held in Innsbruck/Igls primarily considered a new compound with antiparkinsonian properties unrelated to any of the conventional drugs – the diphenylpiperidine compound budipine. The present volume records the experimental and clinical data on budipine presented at the meeting. The editors felt especially encouraged by the authorative biochemical, physiological, clinical and therapeutic reviews that were given at this symposium by outstanding experts in the field of extrapyramidal disorders. Their inclusion should make this volume a valuable source of information on a variety of current research aspects into Parkinson's disease.

Innsbruck/London, October 1984 The Editors

Contents

James Parkinson and the Treatment of His Disease. G. Stern 1

Biochemical Neuropathology: A New Approach to Parkinson's Disease. Y. Agid. With 1 Table 4

Clinical Neurophysiology of Parkinson's Disease. P.J. Delwaide and M. Gonce 13

Dopaminergic Agonists and Receptors: Their Relevance to the Treatment of Parkinson's Disease. D.B. Calne and T.A. Larsen. With 2 Tables 22

Brain Dopaminergic Receptor Changes in Parkinson's Disease in Relation to the Disease Process and Treatment. U.K. Rinne, J.O. Rinne, J.K. Rinne and K. Laakso. With 7 Tables 31

Analysis of Akinesia. H. Narabayashi 40

Observations of Voluntarily Induced Automatic Motor Behaviour in Parkinsonism. J.P.W.F. Lakke and W. van den Burg. With 5 Figures and 1 Table 44

Significance of Mental Disorders in the Long-Term Treatment of Parkinson's Disease. P.-A. Fischer 52

Treatment of Parkinson's Disease in Early and Late Phases. M. Yahr. With 9 Figures and 8 Tables 58

Relevance of Dopaminergic Agonists and Antagonists to Parkinsonian Tremor. S. Ruggieri, A. Denaro, M. Baldassarre, F. Stocchi and A. Agnoli. With 5 Figures 68

Therapeutic Possibilities in On-Off Phenomena in Parkinson's Disease. E. Csanda and M. Tárczy. With 2 Figures and 5 Tables .. 73

Target and Additive Drugs in the Treatment of Parkinson's Disease: A Pilot Study. W. Birkmayer and P. Riederer 79

Pharmacologic Bases of Antiparkinsonian Therapy. H.G. Menge and U. Brand. With 1 Figure and 6 Tables 82

The Pharmacodynamics of Budipine on Central Neurotransmitter Systems. J. Offermeier and J.M. van Rooyen. With 21 Figures ... 93

Biochemical and Pharmacologic Aspects of the Mechanism of Action of Budipine. H. Przuntek and J.-P. Stasch. With 1 Figure and 1 Table 107

Pharmacokinetics and Metabolism of Budipine in Animals and Humans. K. Zech, E. Sturm and G. Ludwig. With 10 Figures 113

Clinical Experience with Budipine in Parkinsonian Therapy. W. Poewe, F. Gerstenbrand and G. Ransmayr. With 4 Figures and 5 Tables 122

Controlled Single-Blind Trial of Budipine. L-Dopa, and Amantadine in 102 Patients with Parkinson's Disease. J. Iizuka. With 12 Figures and 3 Tables 129

Open Clinical Trial of Budipine in 25 Tremor-Dominant Parkinsonian Patients. E. Ketz 140

Effect of Budipine on Parkinsonian Tremor Resistant to Other Antiparkinsonian Medication. G. Ulm. With 13 Figures and 4 Tables 142

Budipine: A New Chemical Substance in the Treatment of Parkinson's Disease. J. Siegfried and R. Fischer. With 3 Tables . . 152

Observations on the Therapeutic Effect of Budipine on Parkinson's Disease. Wang Xinde. With 2 Figures and 1 Table . . . 158

Long-Term Treatment with Budipine. F. Oppel. With 5 Figures and 2 Tables 163

Budipine Treatment of Neuroleptic-Induced Parkinsonism. S. Bien and R. Wörz. With 4 Figures and 5 Tables 169

Randomized Double-Blind Trial of Budipine and Biperiden in Patients with Neuroleptic Extrapyramidal Symptoms. P. König and I. Pass-Kosmath. With 8 Tables 176

The Effect of Budipine on the Rheology of the Blood. E. Ott and H. Lechner. With 4 Tables 182

EMG and Computer Analysis of Patients with Tremor and Rigidity Under Budipine Therapy. S. Krafczyk and H. Hökendorf. With 10 Figures and 2 Tables 185

Vigilance Study Using Electrooculographic and Electroencephalographic Methods After Acute Budipine Administration in Healthy Subjects. K. Schaffler. With 4 Figures and 4 Tables . . 194

Budipine in the Treatment of Extrapyramidal Disorders Not Related to Parkinson's Disease. D. Müller. With 1 Table 203

Subject Index 211

List of Contributors

Agid, Y., Laboratoire de Médecine Expérimentale, C.H.U. Pitié-Salpêtrière, 91, Boulevard de l'Hôpital, F-75634 Paris Cedex 13

Agnoli, A., I Clinica Neurologica, Università degli studi di Roma, Viale Dell'Università, 30, I-00100 Roma

Baldassarre, M., I Clinica Neurologica, Università degli studi di Roma, Viale Dell'Università, 30, I-00100 Roma

Bien, S., Universitätsklinik, Abteilung Neuroradiologie, D-7800 Freiburg

Birkmayer, W., Evangelisches Krankenhaus und Ludwig Boltzmann Institut für Klinische Neurobiologie, Neurochemie-Gruppe, Rossauerlände 37, A-1090 Wien

Brand, U., Byk Gulden Pharmazeutika, Byk-Gulden-Str. 2, D-7750 Konstanz

Calne, D.B., Division of Neurology, Health Sciences Centre Hospital, University of British Columbia, Vancouver, British Columbia, VGT lWS, Canada

Csanda, E., Department of Neurology, Semmelweis Medical University, Balassa u. 6, H-1083 Budapest

Delwaide, P.J., Section of Neurology-Clinical Neurophysiology, University of Liège, B-4020 Liège

Denaro, A., I Clinica Neurologica, Università degli studi di Roma, Viale Dell'Università, 30, I-00100 Roma

Fischer, P.-A., Klinikum der Universität Frankfurt, Abteilung für Neurologie, Schleusenweg 2-16, D-6000 Frankfurt/Main 71

Fischer, R., Byk Gulden Pharmazeutika, Byk-Gulden-Str. 2, D-7750 Konstanz

Gerstenbrand, F., Neurologische Universitätsklinik, Anichstr. 35, A-6020 Innsbruck

Gonce, M., Section of Neurology-Clinical Neurophysiology, University of Liège, B-4020 Liège

Hökendorf, H., Ostertal Klinik, Neurologische Rehabilitationsklinik, Am Bosenberg 61, D-6690 St. Wendel/Saar

Iizuka, J., Zentralkrankenhaus der Deutschen Bundeswehr, Abteilung XII, D-5400 Koblenz

Ketz, E., Neurologische Klinik, Kantonsspital, Rorschacher Str. 95, CH-9007 St. Gallen

König, P., Landesnervenkrankenhaus Valduna, Valdunastr. 16, A-6830 Rankweil

Krafczyk, S., Alfried Krupp Krankenhaus, Neurologische Klinik, Alfried-Krupp-Str. 21, D-4300 Essen 1

Laakso, K., Department of Neurology, University of Turku, SF-20520 Turku 52

Lakke, J.P.W.F., Department of Neurology, University Hospital, P.O. Box 30.001, NL-9700 RB Groningen

Larsen, T.A., Division of Neurology, Health Sciences Centre Hospital, University of British Columbia, Vancouver, British Columbia, VGT 1WS, Canada

Lechner, H., Universitäts-Nervenklinik, Auenbrugger Platz 22, A-8036 Graz

Ludwig, G., Byk Gulden Pharmazeutika, Byk-Gulden-Str. 2, D-7750 Konstanz

Menge, H.G., Byk Gulden Pharmazeutika, Byk-Gulden-Str. 2, D-7750 Konstanz

Müller, D., Neurochirurgische Universitätsklinik Eppendorf, Martinistr. 52, D-2000 Hamburg 20

Narabayashi, H., Juntendo University School of Medicine, 2-1-1 Hongo, Bunkyo-ku, Tokyo 113, Japan

Offermeier, J., Department of Pharmacology, Potchefstroom University, 2520 Potchefstroom, South Africa

Oppel, F., Universitätsklinikum Steglitz, Neurochirurgische/Neurologische Klinik und Poliklinik, Hindenburgdamm 30, D-1000 Berlin 45

Ott, E., Universitäts-Nervenklinik, Auenbrugger Platz 22, A-8036 Graz

Pass-Kosmath, I., Landesnervenkrankenhaus Valduna, Valdunastr. 16, A-6830 Rankweil

Poewe, W., Neurologische Universitätsklinik, Anichstr. 35, A-6020 Innsbruck

Przuntek, H., Neurologische Universitätsklinik und Poliklinik im Kopfklinikum, Josef-Schneider-Str. 11, D-8700 Würzburg

Ransmayr, G., Neurologische Universitätsklinik, Anichstr. 35, A-6020 Innsbruck

Riederer, P., Evangelisches Krankenhaus und Ludwig Boltzmann Institut für Klinische Neurobiologie, Neurochemie-Gruppe, Rossauerlände 37, A-1090 Wien

Rinne, U.K., Department of Neurology, University of Turku, SF-20520 Turku 52

Rinne, J.K., Department of Neurology, University of Turku, SF-20520 Turku 52

Rinne, J.O., Department of Neurology, University of Turku, SF-20520 Turku 52

Ruggieri, S., I Clinica Neurologica, Università degli studi di Roma, Viale Dell'Università, 30, I-00100 Roma

Schaffler, K., Labor für Elektrophysiologie, Albert-Schweitzer-Str. 66, D-8000 München 83

Siegfried, J., Neurochirurgische Universitätsklinik, Rämistr. 100, CH-8091 Zürich

Stasch, J.-P., Neurologische Universitätsklinik und Poliklinik im Kopfklinikum, Josef-Schneider-Str. 11, D-8700 Würzburg

Stern, G., Department of Neurology, University College Hospital, Gower Street, GB-London WC1

Stocchi, F., I Clinica Neurologicyà, Università degli studi di Roma, Viale Dell'Università, 30, I-00100 Roma

Sturm, E., Byk Gulden Pharmazeutika, Byk-Gulden-Str. 2, D-7750 Konstanz

Tárczy, M., Department of Neurology, Semmelweis Medical University, Balassa u. 6, H-1083 Budapest

Ulm, G., Paracelsus Elena Klinik, Klinikstr. 16, D-3500 Kassel-Harleshausen

van den Burg, W., Department of Neurology, University Hospital, P.O. Box 30.001, NL-9700 RB Groningen

van Rooyen, J.M., Department of Pharmacology, Potchefstroom University, 2520 Potchefstroom, South Africa

Wörz, R., Waldparkstr. 20, D-7525 Bad Schönborn 1

Xinde, Wang, Department of Neurology, Hospital Beijing, Beijing, China

Yahr, M.-D., Department of Neurology, Mount Sinai Hospital of Medicine of the City University of New York, New York, NY 10029, USA

Zech, K., Byk Gulden Pharmazeutika, Byk-Gulden-Str. 2, D-7750 Konstanz

James Parkinson and the Treatment of His Disease

G. Stern

The poet George Santayana (1863-1952) wrote that those who cannot remember the past are condemned to repeat it. At this meeting when we are considering contemporary knowledge of the pathophysiology of Parkinsons's disease its treatment and the assessment of a new agent, it may give us a salutory sense of historical perspective to consider the past so that we can judge more critically recent achievements. James Parkinson (1755-1824) was a general practitioner who lived and worked in Hoxton Square, now in central London, but in the late eighteenth century, a small hamlet to the east of the City. A prolific writer on matters medical, scientific and sociological, his biography has been well documented by others, but unfortunately no known portrait has survived. We are dependent upon the observations of a fellow geologist, Dr. Mantell, who wrote "Mr. Parkinson was rather below middle stature, with an energetic, intelligent and pleasing expression of countenance and of mild and courteous manner; readily imparting information either in his favourite science or on professional subjects". His many writings ranged from the regulation of madhouses to observations on the bringing up of children, but the best known of course is his *Essay on the Shaking Palsy* pulished in 1817 from Paternoster Row, London, which justly gave him eponymous but posthumous international fame.

Too modest to accept originality, James Parkinson in his first chapter respectfully refers to the observations of previous writers including Galen, Sylvius de la Boe, Juncker and the "celebrated Cullen with his accustomed accuracy". While earlier writers lacked Parkinson's perspicacity in appreciating the distinctness of his syndrome, previous physicians were probably aware at least of the similar concomitants of advancing age. My attention has been drawn to an Egyptian hieroglyphic about 5000 years old: "the god Ra created all things on earth including man and woman in human form. When he grew old he shook with the palsy and he dribbled at the mouth". Dr. H. Smith, Professor of Egyptology at University College, London, has confirmed the existence of this hieroglyphic and is currently rescrutinising the authenticity and priority of the ancient Egyptian physicians and also the treatments that they prescribed.

Following James Parkinson's essay, during the nineteenth and early twentieth centuries a large number of treatments were introduced, a hallmark of which was usually uncritical enthusiasm rather than demonstrable efficacy or scientific foundation. Among many the calabar bean, parathormone and heavy metals will be recalled; one of the lesser known of Gowers' mixtures contained arsenic, heroin and cocaine which should have reliably alleviated anxiety enhancement of static tremor.

A variety of physical treatments were also prescribed from special diets to vigorous exercises, aperients and purges – James Parkinson certainly felt the last to be important and recommended moderate doses of calomel with the occasional help of Epsom salts, and also in the custom of his time advised that blood should first be taken from the upper part of the neck after which vesicatories should be applied to the same part and a "purulent discharge obtained by appropriate use of the Sabine Liniment". However, even in the matter of therapeutics he retained critical judgement: "until we are better informed respecting the nature of this disease, the employment of internal medicines is scarcely warrantable; unless the malady should point out some remedy, the trial of which rational hope might authorise".

The first known reference to the use of anticholinergic drugs was recorded in 1867 when Ordenstein, a pupil of Charcot mentioned in his thesis *"une dernière expérience thérapeutique. Depuis le commencement de ce mois M. Charcot préscrit chaque jour deux ou trois granules d'hyoscyamine, d'environ un milligramme chacun. Ce médicament a pu procurer quelques heures de repos à plusieurs malades".* Why Charcot made this shrewd therapeutic suggestion remains a matter of speculation, but it is probable that he knew of the effects of the plant Atropa belladonna described by Greek and Roman writers and was familiar with the observation that among the symptoms of poisoning was dryness of the mouth and perhaps therefore recommended the plant extract for the use of excessive salivation and subsequently for tremor and rigidity.

Leaving medicinal treatments, one or two controversial physical methods might be mentioned. All clinicians are aware of the beneficial effects of vibration. Henry Head, the distinguished English neurologist, who was afflicted with Parkinson's disease, claimed that he felt better when he was riding in a bumpy carriage. Robert Schwab at the instigation of one of his patients who only felt well when he was at work driving a tram, designed a vibrating chair. Other patients have made similar discoveries for which there is some scientific support in that vibratory stimuli are known to alter the sensitivity of muscle spindles. Electroconvulsive therapy (ECT) has been mentioned in the literature from time to time with the suggestion that spectacular improvement may occur not only in associated depression, but also in Parkinsonian disabilities. Recently there was also a suggestion that ECT might be of value in the management of the on-off phenomenon. Our experience indicates that the latter is not so, but for those parkinsonian patients who remain profoundly incapacitated by depression despite conventional medication we agree that it is still reasonable to consider ECT.

With respect to the modern era of levodopa therapy, it will be recalled that the seminal observation followed the introduction to western medicine of an extract from the plant *Rauwolfia Serpentina* used in ancient Hindu medicine for over 3000 years in the treatment of lunacy and agitation following certain snake bites. Within a year of its use in modern psychiatry in 1955, it was shown that a minority of patients who had taken *Rauwolfia* continuously for a year developed what we now call drug-induced parkinsonism. It was the elucidation of this phenomenon and the ameliorating effect of levodopa which triggered the explosion of scientific interest into the pathophysiology and treatment of Parkinson's disease from which we now benefit.

Of recent developments perhaps the most intriguing stems from the demonstration by Perlow and his colleagues in 1979 that it was possible to transplant foetal rat substantia nigra into the lateral ventricles of rats and that the cells were active in producing dopamine and serotonin. In a further paper the authors amplify their findings, showing that grafts may remain active for up to 10 months (a long time in the life span of the laboratory rat) and that the graft partially corrected the consequences of 6-hydroxydopamine-induced damage to the substantia nigra and thereby reduced abnormal turning activity. Colleagues at the Karolinska Institute in Stockholm have now initiated investigations in humans and have transplanted cells from the subadrenal gland into the lateral ventricle of the same patient in the hope that the cells will survive and generate central dopamine. It is too soon to determine whether this strategy has practical value, but we await the results of further studies with great interest.

James Parkinson, clearly well aware of the dangers of premature judgement in his characteristically modest apologia for publishing his essay, wrote: "the writer will repine at no censure which the precipitate publication of mere conjectural suggestion may incur: but shall think himself fully rewarded by having excited the attention of those who may point out the most appropriate means of relieving a tedious and most distressing malady". I feel sure that he would have approved and enjoyed this meeting and, like me, would look forward with equal pleasure to the remaining papers concerning his illness and a critical evaluation of a new medication to relieve the disease.

Biochemical Neuropathology: A New Approach to Parkinson's Disease

Y. Agid

Introduction

In Parkinson's disease a dopamine deficiency is observed in the striatum, cortical and limbic areas as well as the hypothalamus. Thus, not only the nigrostriatal, but also the mesocorticolimbic and hypothalamic dopaminergic innervation is affected. Degeneration of the respective dopaminergic systems may explain some aspects of motor, psychic and endocrine abnormalities. However, other neuronal systems are most likely involved in the genesis of parkinsonian symptoms. Indeed, though some systems do not seem to be affected in this disease (i.e. histidine decarboxylase, TRH containing neurons) many others are partially damaged, as suggested by the decrease in transmitter content restricted to some areas. This is the case for noradrenergic, serotonergic as well as some peptidergic systems (methionine-enkephalin, CCK-8, substance P), the clinical implication of which seems unclear at present. The postulated degeneration of the cholinergic innominatocortical system might be of great interest since it has been related to the occurrence of dementia in patients. Progressive understanding of the biochemical state of the parkinsonian brain, and correlations between the biochemical defects and the occurrence of the various symptoms should help in improving the therapeutic follow-up of patients.

The anatomico-clinical method may be complemented by biochemical correlation. Biochemical neuropathology aims to study the distribution of diverse biochemical parameters such as neurotransmitters, peptides and receptor sites in the human brain, post-mortem. The results thus obtained are compared with histological data and clinical observations obtained in vivo. Conditions post-mortem (time elapsing between death and the autopsy, low temperature storage or anatomical specimens) are limiting factors, but can nonetheless be controlled. Circumstances immediately preceding death (therapy, agony) may, on the other hand, modify the results and must be carefully analysed. This experimental approach is particularly valuable in the study of degenerative diseases, of which Parkinson's disease is the classic example.

Although the degeneration of the dopaminergic nigrostriatal pathway is characteristic in Parkinson's disease, its complex symptomatology (mainly motor manifestations, but also mental and endocrine) suggests that other neuron systems are involved. The aim of this review is to show that not only the dopaminergic nigrostriatal system is affected, but also a large number of nondopaminergic circuits, while others remain intact (Table 1).

Table 1. Neuromediators, enzyme activities, and peptides in brains of parkinsonian patients

	Basal ganglia				Mesencephalon		Hypo-thalamus	Cerebral cortex		References
	CN	Put	Pal E	Acc	SNC	VTA		CXF	Hipp	
Neuromediators										
Dopamine	↘	↘	↘	↘	↘	↘	↘	↘	↘	Hornykiewicz (1966, 1980), Javoy-Agid et al. (1982a), Scatton et al. (1982), Ploska et al. (1982)
Noradrenaline	N. D.	N. D.	↘	↘	↘	↘	=	↘	↘	Taquet et al. (1982), Hornykiewicz (1980), Scatton et al. (1983)
Serotonin	↘				↘			↘	↘	Scatton et al. (1983)
Enzyme activity										
CAT	=	=	=	=	=	=	=	↘	↘	Ruberg et al. (1982), Dubois et al. (1983)
GAD	=	=	=	=	=	=	=	↘	↘	Javoy-Agid et al. (1984c), Montfort et al. (to be published)
HD	=				=		=	=		Garborg et al. (1983)
Peptide										
Met-enkephalin	=	↘	↘	=	↘	↘	=	=	=	Taquet et al. (1982), Ploska et al. (1982)
CCK-8	=	=	=	=	↘	=		=	=	Studler et al. (1982)
SP	=	=	↘	=	↘	=		=	=	Mauborgne et al. (1983)
TRH	=	=	=	=	=	=	=	=	=	Javoy-Agid et al. (1984c)

Normal (=) and decreased (↘) content in parkinsonian brains as compared with controls. Abbreviations: *CAT* choline acetyltransferase; *CCK*-8 cholecystokinin-8; *GAD* glutamic acid decarboxylase; *HD* histidine decarboxylase; *SP* substance P; *TRH* thyrotropin-releasing hormone; *Acc* nucleus accumbens; *CN* caudate nucleus; *CXF* frontal cortex (Brodmann area 9); *Hipp* hippocampus; *Pal E* external pallidum; *Put* putamen; *SNC* substantia nigra (pars compacta); *VAT* ventrotegmental area; N.D. = not detectable.

Degeneration of the Dopaminergic Systems

Since Hornykiewicz (1966) showed the presence of degeneration of the nigrostriatal dopaminergic pathway (with a reduction in the concentration of dopamine in the substantia nigra and striatum), we may consider that all the dopaminergic systems in the brain are damaged, and in particular the mesocorticolimbic and hypothalamic systems.

Degeneration of the Dopaminergic Mesocorticolimbic System

The dopamine detected in the mesencephalon of control brains (ventrotegmental area and substantia nigra pars compacta) is contained in the dopaminergic neurons.

1. In humans, the distribution of the amine (Taquet et al. 1982) corresponds exactly to the location of nerve cell bodies, be they pigmented, containing melanine (whose origin is catecholaminergic (Bazelon et al. 1967) or marked by an anti-tyrosine hydroxylase antibody (Gaspar et al. 1983).

2. The results agree with biochemical and immunocytochemical data obtained in animals (Lindvall et al. 1974).

In rats, the neurons in the ventrotegmental area are known to innervate the prefrontal cortex preferentially as well as the principal limbic areas (cingular and entorhinal cortex, amygdala, nucleus accumbens, septum) while neurons of nigral origin project towards the striatum (Porrino et al. 1982). The presence of dopamine and its principal metabolites [dihydroxyphenylacetic acid (DOPAC) and homovanillic acid (HVA)] not only in the striatum, but also in the limbic and cortical areas of the human brain (Hornykiewicz 1980; Javoy-Agid et al. 1982a; Scatton et al. 1983) confirms results obtained from animal experiments using histofluorescence (Lindvall et al. 1974; Berger et al. 1980) indicating the presence of dopaminergic nerve endings. Such is the main evidence for the presence of this dopaminergic system in humans.

In Parkinson's disease, the reduction in the concentration of dopamine in the ventral mesencephalon (substantia nigra and ventrotegmental area) (Javoy-Agid et al. 1982a) suggests the destruction of the dopaminergic neurons in this area. The results are in agreement with a reduction in the number of neurons stained by an anti-tyrosine hydroxylase antibody in the nucleus paranigralis (Javoy-Agid et al. 1982a, 1984). The corresponding dopaminergic nerve endings are probably reduced in number in the cerebral cortex and limbic areas as suggested by the reduction in dopamine, DOPAC and HVA levels in projection sites (Hornykiewicz 1980; Javoy-Agid et al. 1982a; Scatton et al. 1982, 1983). The mesocorticolimbic dopaminergic system is thus damaged in this condition. The effects differ from those in the nigrostriatal pathway in three essential respects:

1. The nigrostriatal system is damaged preferentially. Thus, the reduction in the level of dopamine is of the order of 80% in the nigrostriatal complex, while in the ventrotegmental area and the cerebral cortex it only reaches 50%-60% (Javoy-Agid et al. 1982a; Scatton et al. 1982, 1983).

2. Compensatory hyperactivity of the dopaminergic neurons is observed in the nigrostriatal system and not in the mesocorticolimbic system. This is witnessed by an increase in the HVA: dopamine ratio detected in the caudate nucleus, but is not the case in the hippocampus or in the frontal cortex (Scatton et al. 1983; Javoy-Agid et al.

1984c). The resulting increase in the rate of amine turnover in the striatum probably reflects the hyperactivity of the remaining neurons, as has demonstrated following partial destruction of the nigrostriatal pathway in rats (Agid et al. 1973). This being the case, why is the phenomenon not observed in the mesocorticolimbic system? Perhaps the number of dopaminergic neurons belonging to the corticolimbic system is not sufficiently reduced? Since the HVA/dopamine ratio does not differ in the cortical areas of normal and parkinsonian brains (while this ratio increases threefold in the caudate nucleus of parkinsonian subjects) it may be that the corticolimbic neurons are already fully activated in the physiological (control) situation. It is probable that the activity of these neurons cannot be further increased.

3. The dopaminergic receptor sites (measured using spiperone ^{3}H) are present in increased numbers in the striatum, but not in the limbic structures such as the nucleus accumbens (Bokobza et al. 1984). An increase in the number of striatal dopaminergic receptors must be involved (Bokabza et al. 1984; Rinne et al. 1980) since the affinity constant does not differ between normal and parkinsonian subjects. This hypersensitivity of the nigrostriatal dopaminergic receptors in cases of Parkinson's disease confirms results obtained in animals following dopaminergic denervation (Feuerstein et al. 1981). It is not however observed in the nucleus accumbens, contrary to the findings of Rinne et al. (1980), but in agreement with results obtained from animal experiments (Tassin et al., to be published).

There is, therefore, a difference in the control of the nigrostriatal and mesocorticolimbic systems in Parkinson's disease:

1. When dopaminergic denervation remains moderate (as in the case of the nucleus accumbens) the hyperactivity of the remaining neurons is sufficient to reestablish normal dopaminergic transmission.

2. When dopaminergic denervation becomes too great (as in the case of the striatum) the primary presynaptic compensatory system is overwhelmed and necessitates a secondary compensatory system being brought into play; it is postsynaptic in this case, based on the hypersensitivity of the dopaminergic receptor sites.

Degeneration of the Dopaminergic Systems in the Hypothalamus

Several observations, immunocytochemical in animals (Lindvall et al. 1974), and biochemical in humans (Langevin and Iversen 1980) suggest the presence of dopaminergic neurons in the hypothalamus. In Parkinson's disease, dopamine concentrations are reduced by approximately 50% in this structure (Javoy-Agid et al. 1984a, c) while other factors – the concentration of noradrenaline, adrenaline, methionine-enkephalin, TRH, β-endorphin, γ-lipotropin and other pro-opiomelanocortin derivatives, and the enzymatic activity of glutamic acid decarboxylase and choline acetyltransferase – are no different from control values (Javoy-Agid et al. 1984c). This fall in dopamine levels suggests the selective degeneration of dopaminergic neurons, leaving other systems of neurons unaffected.

Since a nigrohypothalamic dopaminergic pathway has been described in animals (Kizer et al. 1976), this local deficiency of dopamine in Parkinson's disease may be due, at least partly, to the destruction of the afferent mesencephalic dopaminergic neurons. The well-established presence of intrahypothalamic dopaminergic systems in animals and the presence of Lewy bodies in the parkinsonian hypothalamus (Langston and Forno 1978) on the other hand, suggests the destruction of the intrahypothalamic dopaminergic neurons. At all events, the intrahypothalamic dopamine deficiency may play a role in the classic endocrine anomalies found in these patients (Javoy-Agid et al. 1984 a, c).

Degeneration of Nondopaminergic Systems

Among the other systems of neurons, some appear intact and others affected in a diffuse or localised manner. Thus histamine and TRH neurons appear to be unaffected since histidinedecarboxylase activity (enzyme specific to histamine synthesis) (Garbag et al. 1983) and concentrations of TRH (Javoy-Agid et al. 1984 b) are normal in parkinsonian brains. They are therefore only involved to a small extent, if at all, in producing symptoms; this does not however exclude the fact that there may be metabolic disorders in neurons which are otherwise intact. Other systems are partially damaged, as witnessed by biochemical deficiencies localised in certain structures. As a rule, the lowering of the concentration of neurochemical mediators involves the destruction of the corresponding neurons, but metabolic modification is not excluded, be it due to massive dopaminergic degeneration or not. Here is a brief outline of a few examples.

Noradrenergic Systems

The noradrenaline content of the brain is reduced in Parkinson's disease (Scatton et al. 1983). Most ascending noradrenergic systems are damaged, in keeping with the well-established lesion in the locus coeruleus (main origin of noradrenergic neurons whose target is the telencephalon), yet noradrenaline levels are normal in the hypothalamus (Javoy-Agid et al. 1984 a, c). The same is true of dopamine β-hydroxylase activity in the noradrenergic nuclei of the medulla oblongata (Kopp et al. 1982). As the ventral noradrenergic system is, at least in part, unaffected, we may conclude that Parkinson's disease is not characterised by a generalised catecholaminergic deficiency.

Serotonergic Systems

Damage to the serotonergic systems is not ubiquitous and homogeneous, particularly in the cerebral cortex (Scatton et al. 1983), while the serotonergic neurons in the caudate nucleus and hippocampus are greatly affected, as witnessed by a reduction of around 50% in serotonin levels in these structures. The significance of this central serotonergic deficiency in clinical terms is not known; it suggests the partial destruction of the sites in which the serotonergic neurons are presumed to originate, i.e. the nuclei of the raphe.

Cholinergic Systems

The cholinergic systems are normal in Parkinson's disease, particularly in the basal ganglia (Gaspar et al. 1980), except in the cerebral cortex and hippocampus where choline acetyltransferase activity (CAT: enzyme specific to acetylcholine synthesis and taken as an index of the number of cholinergic neurons present) is reduced by approximately 50% (Ruberg et al. 1982). This reduction in CAT activity in the cortical areas is analogous to that which is observed in Alzheimer's disease (Rossor et al. 1982). It is accompanied by hypersensitivity in the corresponding muscarinic receptors, related on one hand to the ante-mortem administration of anticholinergic agents and on the other to cortical cholinergic denervation (Ruberg et al. 1982; Dubois et al. 1983). Interestingly, the reduction in CAT levels is found to correspond to the reduction in the intellectual capacities of parkinsonian subjects (Ruberg et al. 1982; Dubois et al. 1983; Agid et al. 1984), suggesting that the cortical cholinergic deficiency may play an important role in the genesis of dementia. Inversely, these cholinergic neurons might play a role in memory function. Evidence for a reduction in CAT activity in the substantia innominata, the site where cholinergic innervation of cortical destination is thought to originate, suggests that the reduction in CAT activity in the cortex is related to damage to the innominatocortical pathway, i.e. a subcorticocortical cholinergic system. These results are important for two reasons:

1. They may relate dementia in Parkinson's disease to Alzheimer's disease. More than 30% of parkinsonian subjects become demented and show morphological alterations characteristic of Alzheimer's disease; they show, in addition, the same central cholinergic deficiency (Agid et al. 1984).

2. If we assume that there is a connection between subcortical cholinergic degeneration and intellectual deterioration in Parkinson's disease (as in Alzheimer's disease), could the origin of dementia be subcortical?

GABAergic Systems

Contrary to certain previous findings, it appears that GAD activity (glutamic acid decarboxylase, used as a marker of GABAergic neurons) is no different in control brains from those of parkinsonian subjects, at least in the deep structures. This is not the case for the frontal cortex and the hippocampus where a reduction in GAD activity of the order of 50% has been observed (Montfort et al., to be published). Consequently the subcortical GABAergic systems appear to be intact. The reduction in GAD activity in the cortical areas may result either from a loss of neurons or from nonspecific lesions related to ante-mortem conditions and particularly to chronic cerebrovascular insufficiency (Montfort et al., to be published). The clinical and therapeutic implications are unclear.

Peptidergic Systems

The malfunction of the large aminergic and cholinergic pathways is accompanied by local peptidergic deficiency. The spectacular reduction in methionine-enkephalin levels in the ventral mesencephalon (of the order of 80%), and to a lesser degree in the putamen and external pallidum, whereas there is no difference in other cerebral structures compared with control brains (Javoy-Agid et al. 1982 a), could be due to the destruction of afferent met-enkephalinergic neurons in the brain stem (Taquet et al. 1982; Javoy-

Agid et al. 1982a; Ploska et al. 1982). Results obtained in immunocytochemistry (using anti-tyrosine hydroxylase and anti-met-enkephalin antibodies (Gaspar et al. 1983) and the discovery of a significant reduction in the number of opiate receptors in the substantia nigra of parkinsonian subjects (Javoy-Agid et al. 1982 b) suggest that dopaminergic cell bodies are under the direct control of met-enkephalinergic afferences. The functional significance of this met-enkephalinergic deficiency is yet to be elucidated.

The selective reduction in the concentrations of cholecystokinin-8 (CCK-8) solely in the substantia nigra of parkinsonian subjects (Studler et al. 1982) could be explained by its well-established localisation in certain dopaminergic neurons of the ventral mesencephalon, but the absence of any modification of CCK-8 levels in the ventrotegmental area and in the proximity of dopaminergic nerve endings does not support this hypothesis in humans. The role of this limited and local reduction in cerebral CCK-8 remains unexplained.

The demonstration of a reduction in levels of substance P in the substantia nigra (Mauborne et al. 1983) might result from the partial destruction of the striatonigral substance P pathway, but this is only a hypothesis which needs to be confirmed. In all cases these peptidergic alterations are moderate, always localised and difficult to interpret. Do they involve partial neuron degeneration or do they involve secondary functional deficiency following other lesions?

Summary

The predominant deficiency in dopaminergic neurons in the mesostriatal, mesocorticolimbic and hypothalamic systems probably represents the main biochemical characteristic of Parkinson's disease. The incidence of these deficiencies in the symptoms of untreated (akinesia, bradyphrenia), or overtreated patients (involuntary movements, mental disorders due to overdosing) can now be hypothesised. This is not however the case for the other neurochemical systems, some of which are unaffected, such as TRH and histamine containing neurons and others which are partially damaged, such as the noradrenergic, serotonergic and cholinergic pathways and certain peptide systems, e.g. met-enkephalin, CCK-8, substance P. Cortical cholinergic deficiency probably plays a role in memory disorders in parkinsonian subjects and in mental confusion (De Smet et al., 1982) related to excessive anticholinergic treatment (Agid et al. 1984). Briefly, everything takes place as if the systems of long neurons originating in subcortical regions were damaged preferentially, the main alterations being localised in the brain stem and particularly in the ventral mesencephalon.

Acknowledgments. This review summarises data obtained by Doctors B. Bokobza, F. Cesselin, B. Dubois, P. Gaspar, F. Javoy-Agid, A. Mauborgne, J.C. Montfort, A. Ploska, M. Ruberg, B. Scatton, J.M. Studler, H. Taquet who are gratefully acknowledged for their contributions. This work was supported by INSERM (PRC Santé Mentale et Cerveau) and C.N.R.S. (Jeune Equipe).

References

Agid Y, Javoy-Agid F, Glowinski J (1973) Hyperactivity of remaining dopaminergic neurons after partial destruction of the nigro-striatal dopaminergic system in the rat. Nature 245: 150-151

Agid Y, Ruberg M, Dubois B, Javoy-Agid F (1984) Biochemical substrates of mental disturbances in Parkinson's disease. From Proceedings of the VIIth International Symposium on Parkinson Disease, Frankfurt 1982. Adv Neurol 40: 211-218

Bazelon M, Fenichel GM, Randall J (1967) Studies on neuromelanin. A melanin system in the human adult brain stem. Neurology (Minneap) 17: 512-519

Berger B, Tassin JP, Rancurel G, Blanc G (1980) Catecholaminergic innervation of the human cerebral cortex in presenile and senile dementia. Histochemical and biochemical studies. In: Usdin E, Sourkes TL, Youdim MBH (eds) Enzymes and neurotransmitters in mental disease. Wiley, New York, pp 317-327

Bokobza B, Ruberg M, Javoy-Agid F, Agid (1984) ^{3}H-Spiperone binding in putamen and nucleus accumbens of parkinsonian patients. Eur J Pharmacol 99: 167-175

De Smet Y, Ruberg M, Serdaru M, Dubois B, Lhermitte F, Agid Y (1982) Confusion, dementia and anticholinergics in Parkinson's disease. J Neurol Neurosurg Psychiatry 45: 1161-1164

Dubois B, Ruberg M, Javoy-Agid F, Ploska A, Agid Y (1983) A subcortico-cortical cholinergic system is affected in Parkinson's disease. Brain Res 288: 213-218

Feuerstein C, Demenge P, Caron P, Barrette G, Guerin B, Mouchet P (1981) Supersensitivity time course of dopamine antagonist binding after nigro-striatal denervation: Evidences for early and drastic changes in the rat corpus striatum. Brain Res 226: 221-234

Garbarg M, Javoy-Agid F, Schwartz JC, Agid Y (1983) Histidine decarboxylase activity in brains of patients with Parkinson disease. Lancet 1: 74-75

Gaspar P, Javoy-Agid F, Ploska A, Agid Y (1980) Regional distribution of neurotransmitter specific enzymes in the basal ganglia of human brain. J Neurochem 34: 278-283

Gaspar P, Berger B, Gray M, Hamon M, Cesselin F, Vigny A, Javoy-Agid F, Agid Y (to be published) Immunohistochemical study of tyrosine-hydroxylase and methionine-enkephalin in human mesencephalon: distribution and relationships. J Neurol Sci

Hornykiewicz O (1966) Dopamine (3-hydroxytyramine) and brain function. Pharmacol Rev 18: 925-964

Hornykiewicz O (1980) Biochemical abnormalities in some extrastriatal neuronal systems in Parkinson's disease. In: Rinne UK, Kingler M, Stamm G (eds) Parkinson's disease. Current progress, problems and management. Elsevier/North Holland, Amsterdam, pp 109-120

Javoy-Agid F, Taquet H, Berger B, Gaspar P, Morel-Maroger A, Montastruc JL, Scatton B, Ruberg M, Agid Y (1982 a) Relations between dopaminergic and methionine-enkephalinergic systems in control and parkinsonian brains. In Proceedings of the 12th World Congress of Neurology, Kyoto. Excerpta Medica, Amsterdam, pp 188-202

Javoy-Agid F, Ruberg M, Taquet H, Studler JM, Garbarg M, Llorens C, Schwartz JC, Grouselle D, Lloyd K, Raismann R, Agid Y (1982 b) Biochemical neuroanatomy of the human substantia nigra (pars compacta) in normal and in parkinsonian subjects. Adv Neurol 35: 151-164

Javoy-Agid F, Pique L, Ruberg M, Ploska A, Taquet H, Bertagna X, Agid Y (1984 a) Biochemical analysis of hypothalamus in Parkinson disease. Neurolog 34: 672-675

Javoy-Agid F, Grouselle D, Tixier-Vidal A, Agid Y (1984 b) TRH-immunoreactivity in control and parkinsonian brains. Neuropeptides 3: 405-410

Javoy-Agid F, Ruberg M, Taquet H, Bokobza B, Agid Y, Gaspar P, Berger B, N'guyen-Legros J, Alvarez C, Gray F, Escourolle R, Scatton B, Rouquier L (1984 c) Biochemical neuropathology of Parkinson disease. From Proceedings of the VIIth International Symposium on Parkinson disease, Frankfurt 1982. Adv Neurol 40: 189-197

Kizer JS, Palkovits M, Brownstein JM (1976) The projections of the A8, A9, and A10 dopaminergic cell bodies: evidence for a nigral-hypothalamic-median eminence dopaminergic pathway. Brain Res 108: 363-370

Kopp N, Denoroy L, Thomasi M, Gay N, Chazot G, Renaud B (1982) Increase in noradrenaline synthesising enzyme activity in medulla oblongata in Parkinson disease. Acta Neuropathol (Berl) 56: 17-21

Langevin H, Iversen LL (1980) A new method for the microdissection of the human hypothalamus, with mapping of cholinergic, GABA and catecholaminergic systems in twelve nuclei and areas. Brain 103: 623-638

Langston JW, Forno JS (1978) The hypothalamus in Parkinson disease. Ann Neurol 3: 129-133

Lindvall O, Bjorklund A, Moore RY, Stenevi J (1974) Mesencephalic dopamine neurons projecting to the neocortex. Brain Res 81: 325-331

Mauborgne A, Javoy-Agid F, Legrand JC, Agid Y, Cesselin F (1983) Decrease of substance-P-like immunoreactivity in the substantia nigra and pallidum of parkinsonian brains. Brain Res 268: 167-170

Montfort JC, Ploska A, Javoy-Agid F, Haun JJ, Dubois B, Agid Y (to be published) Brain glutamate decarboxylase and "premortem severity index" with especial reference to Parkinson's disease. (submitted for publication)

Ploska A, Taquet H, Javoy-Agid F, Gaspar P, Cesselin F, Berger B, Hamon M, Legrand JC, Agid Y (1982) Dopamine and methionine-enkephalin in human brain. Neurosci Lett 33: 191-196

Porrino LJ, Goldmann Rakic P (1982) Brain stem innervation of prefrontal and anterior cingulate cortex in the rhesus monkey revealed by retrograde transport of HRP. J Comp Neurol 205: 63-76

Rinne UK, Koskinen V, Lonnberg P (1980) Neurotransmitter receptors in the Parkinsonian brain. In: Rinne UK, Klingler M, Stamm G (eds) Parkinson disease. Current progress, problems and management, Elsevier/North Holland, Amsterdam, pp 93-107

Rossor MN, Garrett NJ, Johnson AJ, Mountjoy CQ, Roth M, Iversen LL (1982) A postmortem study of the cholinergic and GABA systems in senile dementia. Brain 105: 313-330

Ruberg M, Ploska A, Javoy-Agid F, Agid Y (1982) Muscarinic binding and choline acetyltransferase activity in parkinsonian subjects with reference to dementia. Brain Res 232: 129-139

Scatton B, Rouquier L, Javoy-Agid F, Agid Y (1982) Dopamine deficiency in the cerebral cortex in Parkinson disease. Neurology (N4) 32: 1039-1041

Scatton B, Javoy-Agid F, Rouquier L, Dubois B, Agid Y (1983) Reduction of cortical dopamine, noradrenaline, serotonin and their metabolites in Parkinson's disease. Brain Res. 275: 321-328

Studler JM, Javoy-Agid F,Cesselin F, Legrand JC, Agid Y (1982) CCK-8 immunoreactivity distribution in human brain: selective decrease in the substantia nigra from parkinsonian patients. Brain Res 243: 176-179

Taquet H, Javoy-Agid F, Cesselin F, Hamon M, Legrand JC, Agid Y (1982) Microtopography of methionine-enkephalin, dopamine and noradrenaline in the ventral mesencephalon of human control and parkinsonian brains. Brain Res 235: 303-314

Tassin JP, Simon H, Glowinski J, Bockaert J (to be published) Modulation of the sensitivity of dopaminergic receptors in the prefrontal cortex and the nucleus accumbens: relationship with locomotor activity. Congress of Psychoneuroendocrinology, Montreal, May 1982. Raven, New York

Clinical Neurophysiology of Parkinson's Disease

P.J. Delwaide and M. Gonce

Introduction

The diagnosis of Parkinson's disease can be made with certainty simply on the basis of a clinical examination. In addition, the lowering of dopamine levels in the basal ganglia has offered solutions to many pathophysiological problems. Therefore, one may wonder what role is left to clinical neurophysiology in this disease.

Electrophysiological studies can be done to improve the accuracy of diagnosis and to add something to our understanding of the pathophysiology of the disease. In fact, there are here complementary objectives. The first is to obtain recordings of the clinical signs to make them objective and to quantify them. In fact, the need for objective measurements is becoming more and more evident as a basis for following the evolution of the disease and for assessing the efficacy of new treatments. The techniques must of course be simple, rapid and well standardized to permit comparisons from individual to individual and from laboratory to laboratory. Certain additional techniques have been proposed that would help in the differential diagnosis and reveal subclinical signs. As the symptomatology is multifactorial, one test alone can not reflect the entire clinical picture.

The second goal is to contribute to our pathophysiological knowledge of the syndrome. In fact, it has not yet been possible to define by what mechanism the lowering of dopamine levels modifies muscular tone and motor functions at large. A series of questions remains to be answered such as which neurons and which circuits are either hyper- or hypoactive. The lack of valid experimental models prevents us from answering these questions by neurophysiological studies performed in animals. It is therefore necessary to collect as much information as possible from patients and to use these data as the basis for pathophysiological interpretation. It is desirable to combine in a single test both accurate assessment of a clinical sign and its pathophysiological analysis. Such a test, although difficult to design, would be ideal and could be named a "correlative test".

A yet, the ambitious goals of clinical neurophysiology in Parkinson's disease have not been entirely achieved. This is probably due, at least in part, to the fact that classical techniques had been developed chiefly for the exploration of spinal mechanisms. However, it is clear that the spinal cord is less affected in Parkinson's disease than in spasticity. It has been necessary to wait for the development of new techniques in order to study the function of areas of the nervous system more evolved than the spinal cord. Recently, techniques have been developed which make it possible to study the control of spinal activity by evoking long loop reflexes (Marsden et al. 1973). Applied to Parkinson's disease these techniques have already provided interesting results. Motoricity is also organized at a higher level and it is in the brain that are found the generators of voluntary acts. Their functions can now be studied, notably by recording the electrical activity on the scalp before a voluntary movement (Kornhuber and Deecke 1965). The results obtained in the study of Parkinson's disease shed some light on disturbances of motor programmes and plans.

Modification of Spinal Activity

Tremor

Tremor is easy to record either by accelerometry or by electromyography (EMG) (Calne and Lader 1969). Accelerometry allows us to measure the frequency and amplitude of the tremor, while EMG can give the sequence of muscular activities in antagonistic muscles. Normally, parkinsonian tremor is of the alternating type, i.e. when an EMG burst is recorded from a given muscle, there is silence in its antagonist and vice versa (Shahani and Young 1976). However, this pattern does not occur exclusively; it is common to observe features which typically point to the coexistence of an intention or postural tremor. Recording the tremor can often aid in recognition of the mixed nature of the abnormal movements (Lance et al. 1963). Sometimes, the rhythm of the extensor is double that of the flexor. A phase shift between different segments of the same limb is also possible. When tremor is dubious, giving the patient mental arithmetic problems can make it more obvious, but it is chiefly vibration which makes its existence clear (Hagbarth and Eklund 1966; Burke et al. 1972). When a vibrator is applied to a limb the bursts are more evident, the amplitude but not the frequency being increased. This is a good example of how a technique can reveal subclinical signs. The excitability of the motor nucleus fluctuates regularly in every cycle of tremor; this can be demonstrated by eliciting the H-reflex at different times in the cycle of tremor (Ekbom et al. 1952). The amplitude during the interval between two bursts clearly increases just before the appearance of the second burst, the changes sometimes reaching 300%.

There is a great deal of evidence indicating that tremor is generated in the brain (Albe-Fessard et al. 1966; Jasper and Bertrand 1966). Suffice it to recall the results of stereotactic surgery or the fact that cutting the dorsal roots does not stop tremor (Pollock and Davis 1930). However, an old debate, which is not yet closed, concerns the role of peripheral afferents in the regulation of tremor. Various experiments indicate that, in fact, they intervene. First, the vibration which stimulates the muscle spindles enhances the amplitude of the bursts (as already discussed). Second, local anaesthesia of the motor nerve reduces tremor and thirdly, ischaemia (which blocks conduction first in larger diameter nerve fibres) abolishes the tremor at its inception. Moreover, a muscle twitch elicited by electrical stimulation provokes a halt in rhythmic activity lasting 200-250 ms, whatever the timing of the stimulation in the cycle of tremor (Rondot and Bathien 1978). Certain authors, in comparing parkinsonian tremor and clonus, have claimed that both share many characteristics (Findley et al. 1981). It can be concluded that tremor, which is undoubtedly caused by a central dysfunction, can be triggered and modulated by peripheral kinesthetic inputs. These latter can be used to demonstrate tremor and its characteristics in cases where its existence is in doubt. A strict separation of peripheral and central functions is untenable since these two factors act together.

Rigidity

To record and study rigidity, techniques which have proved useful in the exploration of spasticity were first used. This approach was logical, even more so because in sectioning the dorsal roots rigidity was abolished which in itself suggested the influence of the myostatic reflex arc (Pollock and Davis 1930). Thus, it was found that tendon jerks are more or less normal, indicating that the dynamic response to stretch is not increased (McLellan 1981).

The H-reflex and the H max : M max ratio are also normal, making it clear that the excitability of the motor nuclei is not enhanced (Angel et Hoffmann 1963; Dietrickson 1971; Delwaide 1971). Rigidity results from the progressive recruitment of new motor units as the muscle is stretched with progressive fallout of previously recruited units. The muscle is usually silent at rest, but EMG activity resumes if the muscle is stretched by passive joint movement of more than 5°-10°. Injection of the muscle belly with procaine abolishes this response, confirming that it is induced by the activation of sensory receptors in the limb. The static response which increases in proportion to muscle length is enhanced (Andrews et al. 1972).

The tonic vibration reflex is usually regarded as normal when measured without interference of tremor. However, voluntary control of it is usually reduced (Hagbarth and Eklund 1968). Some authors hold contradictory views. McLellan (1973) for example, believes that the tonic vibration reflex correlates positively with the severity of rigidity. The vibratory inhibition of the monosynaptic reflex which reflects presynaptic inhibition is normal (Delwaide 1971). The silent period which is attributed to withdrawal of spindle input can be considered, despite contradictory reports, as normal in both latency and duration (Angel et al. 1966; Higgins et al. 1971; Dietrickson 1971).

These results obtained with classical methods do not argue in favour of an increased excitability of the myostatic arc. More direct evidence has been anticipated from newer techniques, notably the direct recording of IA fibres (Hagbarth and Vallbo 1967). These recordings, known as microneurography, have shown more spontaneous afferent activity in parkinsonian patients than in relaxed subjects. However, this afferent activity is in proportion to the amount of background EMG activity in the muscles themselves so that no firm conclusion can be drawn regarding the role of the gamma system (Wallin et al. 1973; Burke et al. 1977). Taken together these results indicate that even when an intense rigidity exists the excitability of the myostatic reflex arc is not enhanced.

If the stretch reflex is not modified, what then are the disturbances at the spinal level? The shortening reaction gives us some idea. This shortening reaction is abnormally increased in parkinsonian patients (Rondot and Bathien 1981). The tonic component which is variable in normal subjects is on the contrary marked in parkinsonian patients. It has been shown that this reaction, which is not velocity dependent, is due to joint afferents (Katz and Rondot 1978) and not to IA afferents. This finding suggests that the reactivity of the interneurons mediating this effect is modified.

Other examples of modified interneuronal reactivity can be given. For example, certain very consistent abnormalities were found in Parkinson's disease in studying the reflexes of exteroceptive origin (Delwaide et al. 1974). These reflexes are more easily evoked than normally both in the upper and the lower limbs. This observation holds particularly true for the first component of the EMG response. Moreover, stimulation of the sural nerve induces a less marked suppression of voluntary activity in the leg muscle than normal. These discrepancies can be explained by postulating an abnormal excitability of certain interneurons linking the peripheral exteroceptive afferences and the motoneurons. During the study of exteroceptive reflexes, another fact was found noteworthy: electrical stimulation of a sensory or mixed nerve frequently evoked contemporary EMG responses in antagonistic muscles. This cocontraction which could also reflect abnormalities in interneuronal network has prompted us to make a more refined study of a particuliar interneuron, the IA interneuron, responsible for the reciprocal inhibition. Experiments in which the peroneal nerve is anaesthetized give indirect information concerning its function at the spinal level. After this procedure, the amplitude of the soleus H-reflex increases remarkedly, suggesting that afferences coming from antagonistic muscles induce a tonic and powerful inhibition on the soleus motor nucleus (Rondot and Bathien 1981). With the technique described by Tanaka (1974), we were able to show that the activity of the IA interneuron is not reduced but is in fact enhanced in parkinsonian patients. The same holds true in the upper limb (Day et al. 1981). This is a new example of the excitability of a given interneuron being modified in Parkinson's disease. As one can not explain the cocontraction observed during the study of exteroceptive reflexes on the basis of a reduced activity of the IA interneurons, other modifications in the functioning of interneurons must be postulated. At the present time, changes in the excitability of the interneuronal network appear to be the common denominator of spinal dysfunction in Parkinson's disease. This point is interesting because descending catecholaminergic pathways are known.

Akinesia

There is no specific technique which explores akinesia at the spinal level. It can only be said that the motoneurons can easily be stimulated when exteroceptive afferents are used (Delwaide et al. 1974). It has been shown that there is a good correlation between the degree of akinesia and the lack of habituation of the second component of the blink reflex (Penders and Delwaide 1971). The measure of the habituation index proposed by Penders and Delwaide (1971) as the ratio between the amplitudes of the fifth and first responses of a series recorded at different frequencies could be a good correlative test. In the lower limb, there is also a lack of habituation of the polysynaptic evoked responses.

Long Latency (Long Loop) Reflexes

When a voluntary movement is perturbed, the EMG of the muscles actively involved presents a series of modulations which are secondary to the activation of neuromuscular spindles (Marsden et al. 1973). A first burst M1 occurs within the delay of the stretch reflex. The second burst M2 which has a latency of around twice that of M1, as well as the third burst M3 are possibly attributable to the action of transcortical reflex pathways (Marsden et al. 1978; Lee and Tatton 1978).

Tatton and Lee (1975) were the first to point out the marked increase in the wrist flexor response to wrist extension in parkinsonian subjects, as shown by EMG. The specificity of their results was questioned, however, owing to the fact that this response is often found to be greater in the elderly (Evarts et al. 1979). However, several other groups have confirmed Lee and Tatton's initial findings (Mortimer and Webster 1979; Chan et al. 1979) in other muscle groups of the upper as well as the lower extremities. Tatton et al. (to be published) are now able to report a strong positive correlation between the amplitude of the M2 response and the intensity of rigidity. From the pathogenic viewpoint, given the fact that the motoneuron excitability is normal, it appears that the more marked long loop response indicates an increased gain in the transcortical pathways·M2 recordings therefore may be a correlative test of great interest.

Older experiments support these results in indicating a greater facilitation for H-reflex recovery curves at delays between 100 and 200 ms (phase IV). Owing to its latency, it is possible that this facilitation arises from long loop activation (Zander Olsen and Diamantopoulos 1967; Sax et al. 1977). When monosynaptic reflexes of either the upper or lower extremities are conditioned by stimulation of the sural nerve, facilitation is more evident at a delay consistent with an intervention of long loops conveying exteroceptive afferences (Delwaide and Crenna, to be published). The Jendrassik manoeuvre which most probably also involves a supraspinal relay after conditioning contraction (Delwaide and Toulouse 1981) has so far furnished contradictory results (Hoffmann et al. 1962; Dietrickson 1971; Arrigo et al. 1962; Gassel and Diamantopoulos 1964). However, this manoeuvre clearly facilitates EMG reflex responses of exteroceptive origin (Delwaide et al. 1974).

Marsden et al. (1981) described a mechanism of postural control induced by kinesthetic stimuli in a normal subject. If the arm meets an unexpected perturbation, early automatic EMG responses may be found in distant postural muscles. This response, called the anticipatory postural reflex, is both altered and diminished in Parkinson's disease (Traub et al. 1980). It can thus said that the regulation of motoricity exerted by long loop reflexes is clearly perturbed in Parkinson's disease, no doubt at the supraspinal level. However, the relays of the long loop reflexes are not located in the basal ganglia themselves.

The Generators of Motoricity

To study the generators of motoricity, voluntary movements are mainly used. Ballistic movements, which are not influenced by peripheral feedback, are well suited for this purpose. It has been found that reaction time in parkinsonian patients is not significantly prolonged, indicating that the planning of movement is not fundamentally disturbed (Evarts et al. 1981).

Just as in a normal subject, a ballistic movement in a parkinsonian patient consists of a series of EMG bursts in both the agonist and the antagonist muscles. The normal

EMG pattern is preserved. However, these EMG bursts are inadequate to move the limb to the required point. To reach its goal, many successive patterns of agonist-antagonist EMG bursts are necessary as shown by Hallett and Khoshbin (1980). These authors have therefore suggested that a normal role of the basal ganglia would be to energize the appropriate muscles required to make the movement. The patients cannot initiate sufficient force to complete the movement. As a result, patients with Parkinson's disease are constrained to move slowly towards their objective. The simplicity of the technique used by Hallett and Koshbin deserves to be mentioned; only a simple EMG was needed to reach these conclusions.

Another approach used in the study of motoricity generators is the recording of electrical potentials derived from the scalp during the preparation for movement. A slow negative potential, the *Bereitschaftspotential* or readiness potential, is observed in the precentral area, contralateral to the limb to be moved before a voluntary, but not before a passive movement (Kornhuber and Deecke 1965).

Anomalies of this potential have been described in parkinsonian patients by Deecke et al. (1977) and by Shibasaki et al. (1978), i.e. reduced amplitude and longer duration. Clear differences of potential are found contralaterally on the scalp in patients whose affection is unilateral; in that case, reduction of the potential is observed on the contralateral side of the clinical deficit.

Miscellaneous

The visual (Bodis-Wollner and Yahr 1978; Delwaide et al. 1980) and auditory responses (Gawel et al. 1981) are both delayed and of reduced amplitude in parkinsonian patients. There is a good correlation between these abnormalities and the clinical classification (Hoehn and Yahr Scale).

It must be realized that most techniques now available are directed towards the study of spinal mechanisms. However, decisive progress has recently been made in the exploration of supraspinal motor mechanisms. One would hope that this will lead in the next few years to better clinical evaluation of patients and a better pathophysiological understanding of the Parkinson's disease.

References

Albe-Fessard D, Guiot G, Lamarre Y, Arfel G (1966) Activation of thalamo-cortical projections related to tremorogenic process. In: Purpura D, Yahr M (eds) The thalamus. Columbia University Press, New York, pp 237-254

Andrews CJ, Burke D, Lancet JW (1972) The response to muscle stretch and shortening in parkinsonian rigidity. Brain 95: 795-812

Angel RW, Hoffmann WW (1963) The H reflex in normal, spastic and rigid subjects. Arch Neurol 8: 591-596

Angel RW, Hoffmann WW, Eppler W (1966) Silent period in patients with parkinsonian rigidity. Neurology (Minneap) 16: 529-532

Arrigo A, Cosi V, Savaldi F (1962) Modification du réflexe rotulien au cours de la manoeuvre de Jendrassik chez les sujets parkinsoniens. Electroencephalogr Clin Neurophysiol 22: 98-99

Bodis-Wollner I, Yahr MD (1978) Measurements of visual evoked potentials in Parkinson's disease. Brain 101: 805-813

Burke D, Andrews CJ, Lance JW (1972) Tonic vibration reflex in spasticity, Parkinson's disease and normal subjects. J Neurol Neurosurg Psychiat 35: 477-486

Burke D, Hagbarth KE, Wallin BG (1977) Reflex mechanisms in Parkinsonian rigidity. Scand J Rehabil Med 9: 15-23

Calne DB, Lader MH (1969) Electromyographic studies of tremor using an averaging computer. Electroencephalogr Clin Neurophysiol 26: 86-92

Chan CW, Kearney RE, Melvill Jones G (1979) Tibialis anterior response to sudden ankle displacements in normal and parkinsonian subjects. Brain Res 173: 303-314

Day BL, Marsden CD, Obeso JA, Rothwell JC (1981) Peripheral and central mechanisms of reciprocal inhibition in the human forearm. J Physiol (Lond) 317: 59-60

Deecke L, Englitz HG, Kornhuber HH, Schmitt G (1977) Cerebral potentials preceding voluntary movement in patients with bilateral or unilateral Parkinson akinesia. In: Desmedt JE (ed) Attention, voluntary contraction and event-related cerebral potentials. Prog. clin. Neurophysiol. Karger, Basel (Progress in clinical neurophysiology, vol 1, pp 151-163)

Delwaide PJ (1971) Etude expérimentale de l'hyperréflexie tendineuse en clinique neurologique. Arscia, Bruxelles

Delwaide PJ, Crenna P (to be published) Exteroceptive influences on motoneuronal excitability in man: spinal and supraspinal contributions. In: Desmedt JE (ed) Motor mechanisms in health and disease. Raven, New York

Delwaide PJ, Toulouse P (1981) Facilitation of monosynaptic reflex by voluntary contraction of muscles in remote parts of the body. Mechanisms involved in the Jendrassik manoeuvre. Brain 104: 701-719

Delwaide PJ, Schwab RS, Young RR (1974) Polysynaptic spinal reflexes in Parkinson's disease. Neurology (Minneap) 24: 820-827

Delwaide PJ, Mesraoua B, De Pasqua V (1980) Les potentiels évoqués visuels dans la maladie de Parkinson. Rev Electroencephalogr Neurophysiol Clin 10: 338-342

Dietrickson P (1971) Phasic ankle reflex in spasticity and parkinsonian rigidity: the role of the fusimotor system. Acta Neurol Scand 47: 22-51

Dietrickson P, Langbretson OF, Houland J (1978) Quantitation of tremor in man. In: Desmedt JE (ed) Progress in clinical neurophysiology, vol 5. Karger, Basel, pp 90-94

Ekbom KA, Jernelius B, Kugelberg E (1952) Notes on variations in muscle stretch reflexes in relation to tremor in Parkinsonism. Acta Med Scand 141: 301-304

Evarts EV, Teravainen H, Beuchert BE, Calne DB (1979) Pathophysiology of motor performance in Parkinson's disease. In: Fuxe K, Calne DB (eds) Dopaminergic ergot derivatives and motor function. Pergamon, Oxford, pp 45-49

Evarts EV, Teravainen H, Calne DB (1981) Reaction time in Parkinson's disease. Brain 104: 167-186

Findley LJ, Gresty MA, Halmagyi GM (1981) Tremor, the cogwheel phenomenon and clonus in Parkinson's disease. J Neurol Neurosurg Psychiat 44: 534-546

Gassel MM, Diamantopoulos E (1964) The Jendrassik maneuver. I. The pattern of reinforcement of monosynaptic reflexes in normal subjects and patients with spasticity or rigidity. Neurology (Minneap) 14: 555-560

Gawl MJ, Das P, Vincent S (1981) Visual and auditory evoked responses in Parkinson's disease. In: Rose FC, Capildeo R (eds) Research progress in Parkinson's disease. Pitman, London, pp 138-146

Hagbarth KE, Eklund G (1966) Motor effects of vibratory stimuli in man. In: Granit R (ed) Muscular afferents and motor control. Almqvist and Wiksell, Stockholm (Nobel symposium, vol 1)

Hagbarth KE, Eklund G (1968) The effects of muscle vibration in spasticity, rigidity and cerebellar disorders. J Neurol Neurosurg Psychiat 31: 207-213

Hagbarth KE, Vallbo AB (1967) Mechanoreceptor activity recorded percutaneously with semi-microelectrodes in human peripheral nerves. Acta Physiol Scand 69: 121-122

Hallett M, Khoshbin S (1980) A physiological mechanism of bradykinesia. Brain 103: 301-304

Higgins DC, Haidri NH, Wilbourn AJ (1971) Muscle silent period in Parkinson's disease. J Neurol Neurosurg Psychiatry 34: 508-511

Hoffmann WW, Angel RW, Seil FJ (1962) Reflex reinforcement in parkinsonism. Neurology (Minneap) 12: 440-442

Jasper HH, Bertrand G (1966) Thalamic units involved in somatic sensation and voluntary and involuntary movements in man. In: Purpura D, Yahr M (eds) The thalamus. Columbia University Press, New York, pp 365-390

Katz R, Rondot P (1978) Muscle reaction to passive shortening in normal man. Electroencephalogr Clin Neurophysiol 45: 90-99

Kornhuber HH, Deecke L (1965) Hirnpotentialänderungen bei Willkürbewegungen und passiven Bewegungen des Menschen: Bereitschaftspotential und reafferente Potentiale. Pflügers Arch 284: 1-17

Lance JW, Schwab RS, Peterson EA (1963) Action tremor and the cogwheel phenomenon. Brain 86: 95-110

Lee RG, Tatton WG (1978) Long loop reflexes in man: clinical applications. In: Desmedt JE (ed) Cerebral motor control in man. Long loop mechanisms. Karger, Basel. Progress in clinical neurophysiology, vol 4, pp 320-333

Marsden CD (1982) The mysterious motor function of the basal ganglia: The Robert Wartenberg Lecture. Neurology (Minneap) 32: 514-539

Marsden CD, Merton PA, Morton HB (1973) Is the human stretch reflex cortical rather than spinal? Lancet 1: 759-761

Marsden CD, Merton PA, Morton HB, Adam JE, Hallett M (1978) Automatic and voluntary responses to muscle stretch in man. In: Desmedt JE (ed) Cerebral motor control in man. Long loop mechanisms. Karger, Basel. Progress in clinical neurophysiology, vol, 4, pp 167-177

Marsden CD, Merton PA, Morton HB (1981) Human postural responses. Brain 104: 513-534

McLellan DL (1973) Dynamic spindle reflexes and the rigidity of parkinsonism. J Neurol Neurosurg Psychiatry 36: 342-345

McLellan DL (1981) Rigidity. In: Clifford Rose F, Capildeo R (eds) Research progress in Parkinson's disease. Pitman, London. Progress in neurology series, pp 88-97

Mortimer JA, Webster DD (1979) Evidence for a quantitative association between EMG stretch responses and Parkinsonian rigidity. Brain Res 162: 169-173

Penders CA, Delwaide PJ (1971) Blink reflex studies in patients with Parkinsonism before and during therapy. J Neurol Neurosurg Psychiatry 34: 674-678

Pollock, LJ, Davis L (1930) Muscle tone in parkinsonian states. Arch Neurol Psychiatry 23: 303-311

Rondot P, Bathien N (1978) Pathophysiology of Parkinsonian tremor. A study of the patterns of motor unit discharges. In: Desmedt JE (ed) Physiological tremor, pathological tremor and clonus. Karger, Basel. Progress in clinical neurophysiology, vol 5, pp 138-149

Rondot P, Bathien N (1981) Spinal reflexes in Parkinson's disease: their evolution under treatment. In: Clifford Rose F, Capildeo R (eds) Research progress in Parkinson's disease. Pitman, London. Progress in neurology series, pp 105-110

Sax DS, Johnson TL, Feldman RG (1977) L-Dopa effects on H reflex recovery in Parkinson's disease. Ann Neurol 2: 120-124

Shahani BT, Young RR (1976) Physiological and pharmacological aids in the differential diagnosis of tremor. J Neurol Neurosurg Psychiat 39: 772-783

Shibasaki H, Shima F, Kuroiwa Y (1978) Clinical studies of movement related cortical potential (MP) and the relationship between the dentatorubothalamic pathway and readiness potential (RP). J Neurol 219: 15-25

Tanaka R (1974) Reciprocal Ia inhibition during voluntary movements in man. Brain Res 21: 529-540

Tatton WG, Lee RG (1975) Evidence for abnormal long loop reflexes in rigid Parkinsonian patients. Brain Res 100: 671-676

Tatton WG, Bedingham W, Bruce IC, Blair RDG, Verrier MC (to be published) Physiologic basis of parkinsonian rigidity. In: VIIth International Symposium on Parkinson's Disease, June 27-30, Frankfurt 1982. Raven, New York

Traub MM, Rothwall JC, Marsden CD (1980) Anticipatory postural reflexes in Parkinson's disease and other akinetic-rigid syndromes and in cerebellar ataxia. Brain 103: 393-412

Wallin BG, Hongell A, Hagbarth KE (1973) Recordings from muscle afferents in Parkinsonian rigidity. In: Desmedt JE (ed) New development in electromyography and clinical neurophysiology, vol 3. Karger, Basel, pp 263-272

Zander Olsen P, Diamantopoulos E (1967) Excitability of spinal motoneurons in normal subjects and patients with spasticity, rigidity and cerebellar hypotonia. J Neurol Neurosurg Psychiatry 30: 325-331

Dopaminergic Agonists and Receptors: Their Relevance to the Treatment of Parkinson's Disease

D.B. Calne and T.A. Larsen

Introduction

Following the recognition of dopamine as a neurotransmitter, and the discovery of its depletion in parkinsonism, therapy for this disorder has been focused on achieving increased activation of striatal dopaminergic receptors. Initially, this end was accomplished by administering the precursor of dopamine, levodopa; certain unwanted effects of this amino acid, notably nausea and vomiting, were alleviated by concomitant administration of drugs, (carbidopa and benserazide) that blocked the conversion of levodopa to dopamine by L-aromatic amino acid decarboxylase. These enzyme inhibitors were chosen because they did not readily cross the blood-brain barrier, so they did not impede the formation of dopamine in the striatum.

After the substantial benefits of this approach to treatment became established, it was gradually recognised that important problems in the management of parkinsonism remained, and some of these might even be causally related to prolonged exposure to high doses of levodopa. The major difficulties were declining efficacy, frequent transient fluctuations in the response to therapy (wearing-off and on-off reactions), progressive dyskinesia, and psychiatric symptoms (confusion, hallucinations, delusions). Efforts were therefore made to develop alternative methods of activating dopaminergic receptors. The properties sought in new drugs were improved pharmacokinetics (a more protracted therapeutic action) and enhanced pharmacodynamic selectivity (to achieve a reduction in the side effects of therapy). The approach adopted was an attempt to produce artificial, direct acting agonists of dopamine, by synthesizing a large range of potentially active compounds that were studied in dopaminergic systems in animals, and then tested clinically. However, in addition to the important practical significance of the advent of dopaminergic agonists to therapy, these drugs have had a major impact on the analysis of synaptic mechanisms that utilize dopamine. In particular, they have allowed a subclassification of dopaminerigc receptors to be formulated.

Dopamine Receptors

Because of the known deficiency of dopamine in Parkinson's disease, it has been suggested that receptor abnormalities may also exist in this disorder. Before considering the studies on receptors in Parkinson's disease, the current status of classification of dopaminergic receptor subtypes will be reviewed briefly.

Dopaminergic receptors have been divided into two subgroups: D1 receptors, that are coupled to the enzyme adenylate cyclase, and D2 receptors, that are independent of this enzyme (Kebabian and Calne 1979; Spano et al. 1978). The affinities of various agonist and antagonist drugs support the distinction between D1 and D2 receptor subtypes: D1 receptors respond to dopamine and haloperidol in micromolar concentrations, whereas only nanomolar concentrations of these drugs are needed to activate a D2 receptor. Bromocriptine, lergotrile and lisuride activate D2 receptors, but inhibit D1 receptors.

Recently, further subdivisions of dopaminergic receptors have been proposed. One hypothesis postulates four dopaminergic receptor subtypes. D1 receptors are again distinguished by their linkage to adenylate cyclase, but the former D2 group is divided into three categories. While there is continuing controversy over the evidence for these modifications of the original theory, the division into two subtypes, D1 and D2 receptors, is still the most widely accepted and the most relevant from a clinical viewpoint; both the dopaminergic agonist and antagonist drugs used in neurology and psychiatry appear to mediate their therapeutic action via the D2 receptors. The role of the adenylate cyclase linked D1 receptor in the brain remains obscure.

Although most ergot derivatives affect both D1 and D2 systems, drugs are becoming available that selectively activate either D1 or D2 receptors. An example of the former type is SKF 38393 (Setler et al. 1978) and of the latter type LY-141865 (Tsuruta et al. 1981). Availability of selective drugs for clinical use should help to elucidate the physiological role of the different receptor types, and their involvement in the pathology of degenerative diseases.

In this discussion it is important to recognise that most of the evidence relating to dopaminergic receptors is indirect. Observations are generally made on tissues that bind, specifically, to substances that are active at dopaminergic receptors, as agonists or antagonists; it is inferred that these specific binding sites represent dopaminergic receptors. Abnormalities of dopaminergic receptors have been inferred from ligand binding studies in Parkinson's disease, but the reports are conflicting. Different ligands have been employed and the number of patients studied have usually been small, making the variance of the results large (Table 1).

Reisine et al. (1977) studied specimens from the brains of ten parkinsonian patients. They found spiroperidol binding to be significantly decreased in the caudate compared with control brains. Nonsignificant changes were detected in the putamen and pallidum. All ten patients had received L-dopa or other antiparkinsonian medications (Reisine et al. 1977).

Lee et al. (1978) reported that in five parkinsonian patients, who had never received L-dopa therapy, apomorphine binding was decreased and haloperidol binding increased in the putamen. Apomorphine binding sites were speculated to correlate with presynaptic dopaminergic receptors, whereas haloperidol sites were thought to represent postsynaptic receptors. In four patients who had been treated with L-dopa the haloperidol binding in the putamen did not significantly differ from control values. In the caudate of all patients, apomorphine and haloperidol binding did not significantly differ from controls.

In the study of Quik et al. (1979), no significant changes were detected in spiroperidol binding in parkinsonian caudates irrespective of L-dopa treatment, but decreased binding was claimed to exist in the substantia nigra of material pooled from three patients, two of whom had received L-dopa. In the caudates from six parkinsonian patients treated with L-dopa, Winkler et al. (1980) found decreased spiroperidol binding as compared with controls.

The largest study is that reported by Rinne et al. (1981) who investigated dopaminergic receptors from the brains of 26 parkinsonian patients. In 15 patients, who had never received L-dopa, there was a subgroup of 10 that appeared to have an increase in spiroperidol binding sites in caudate, putamen and pallidum, whereas in 5 patients the binding in caudate and putamen was decreased; in 10 L-dopa-treated patients, a subgroup of 5 had decreased binding in the caudate and putamen, and another 5 patients showed no significant changes compared with controls. The patients with the decreased binding had more severe Parkinson's disease, with a reduction or loss of response to dopaminomimetic medication. The patients in whom the binding was not decreased had displayed a better response to L-dopa, but also had increased dyskinesia. The higher binding occurred in patients who had received neuroleptic therapy.

In summary, there are several reports of changes in neural binding sites in Parkinson's disease, that are thought to correlate with alterations of the number of dopaminergic receptors. The most consistent pattern of change is a decreased number of binding sites in patients treated with L-dopa and possibly in severely affected untreated patients, whereas in most of the parkinsonian patients who had not received L-dopa, both increased and decreased binding has been reported. An increased binding in untreated patients would be in accord with the principal of denervation supersensitivity, which is known to occur at dopaminergic synapses in the brain (Hornykiewicz 1975). In advanced Parkinson's disease, the reported reduction of dopaminergic receptors is in keeping with the concept of pathology extending to result in the declining efficacy of dopaminomimetic therapy, because degeneration starts to affect the caudate neurons postsynaptic to the nigrostrital dopaminergic pathway.

It is interesting that changes have also been reported in extraneural dopamine binding sites in Parkinson's disease. In ten patients who had not received L-dopa, Le Fur et al. (1981) claimed a decrease in spiroperidol binding sites on lymphocytes. L-dopa therapy corrected the abnormality. More information is desirable to confirm or refute this report, but the notion of extrastriatal pathology involving dopamine systems outside the striatum is not new (Hornykiewicz 1980; Eisler et al. 1981).

Dopamine Agonists as Antiparkinsonian Therapy

The first dopamine agonist to be employed in the treatment of parkinsonism, apomorphine, was given before the levodopa era as part of a much broader pragmatic screening programme in which almost every medicament was tried in patients with intractable chronic disease.

Apomorphine

Schwab et al. (1951) gave apomorphine to patients with Parkinson's disease because this drug reduced decerebrate rigidity in animals, and increased resistance to passive

Table 1. Reports of dopaminergic receptor binding studies in different brain regions in patients with Parkinson's disease

		Reisine et al. (1977)	Lee et al. (1978)		Quik et al. (1979)	Winkler et al. (1980)	Rinne et al. (1981)
		Spiroperidol binding	Haloperidol binding	Apomorphine binding	Spiroperidol binding	Spiroperidol binding	Spiroperidol binding
Patients not treated with L-dopa	Caudate		N.S. (4)	N.S. (4)	N.S. (3)		↓ (5) ↑ (10)
	Putamen		↑ (5)	↓ (5)			↓ (5) ↑ (10)
	Pallidum						N.S. (5) ↑ (10)
Patients treated with L-dopa	Caudate	↓ (10)	N.S. (4)		N.S. (4)	↓ (6)	↓ (5) N.S. (5)
	Putamen	N.S. (9)	N.S. (4)				↓ (5) N.S. (5)
	Pallidum	N.S. (5)					N.S. (5) N.S. (5)
	Substantia nigra				↓ (3)		

Abbrevations: ↑ = increased binding compared with controls; ↓ = decreased binding compared with controls; N.S. = no significant difference from controls; number of patients studied in parenthesis.

movement is a cardinal feature of parkinsonism. Schwab et al. had no knowledge that apomorphine was a classical dopaminergic agonist, but they did report clinical benefit in their patients. Cotzias et al. (1972) confirmed this finding, but the drug was inconvenient, because it had to be injected, and it induced excessive emesis and renal toxicity.

N-Propylnoraporphine

Cotzias et al. (1976) studied this derivative of apomorphine in an attempt to find an agent that could be given orally, and would produce fewer unwanted effects. *N*-Propylnoraporphine fulfilled all these needs, but still failed to achieve an adequate therapeutic index for introduction as a routine antiparkinson agent.

Bromocriptine

This drug is the first dopaminergic agonist to be accepted for widespread use in Parkinson's disease. The evolution of bromocriptine is closely linked to the analysis of the physiological mechanisms controlling the release of prolactin from the anterior pituitary. It was found that bromocriptine induced a fall in the plasma concentration of prolactin in animals (Flückiger 1976), and dopamine was identified as the hormone normally secreted by the median eminence of the hypothalamus, from where it is carried via the portal circulation to the anterior pituitary and acts upon receptors of the mammotrophic cells. Bromocriptine elicited its effect by mimicking dopamine at the pituitary, and it was found that this drug simulated dopaminomimetic phenomena at other sites, including the striatum (Corrodi et al. 1973).

Bromocriptine was then given to parkinsonian patients, and efficacy was reported (Calne et al. 1974). There followed numerous confirming reports in which a wide range of different doses of bromocriptine were given to patients with a wide range of different parkinsonian problems. Concomitant therapy underwent a wide variety of differing management, and a wide range of clinical results were reported, from enthusiastic claims of drastic improvement to disdainful cataloguing of adverse reactions. In the light of the experience accumulated over 8 years, some general patterns can be observed from the literature, and reasonably firm conclusions can be drawn.

1. Definite indications for treating patients with bromocriptine include wearing off reactions, severe dyskinesia and early morning dystonia.

2. Definite contraindications include severe confusion, hallucinations, delusions, recent cardiac arrhythmia and recent myocardial ischaemia.

3. The intake of bromocriptine should be increased gradually, starting with a test dose of 1.25 mg and usually reaching an optimal level between 25 and 75 mg/day, the more mildly affected parkinsonian patients and the less robust patients usually being at the lower end of this dose range.

4. As the dose of bromocriptine is raised, the intake of levodopa (including Sinemet and Madopar) should be reduced, ultimately to about 50% of the level preceding treatment with bromocriptine.

5. This whole process is sufficiently complex to justify admission to hospital for many patients, in which case a period of 2-3 weeks of readjustment of the antiparkin-

sonian drug regimen usually suffices. When undertaken on outpatients, the procedure should be more gradual, usually taking 6-8 weeks.

6. The role of bromocriptine in treating early, mild Parkinson's disease, and the value of prolonged low dosage (below 20 mg/day) have not been defined.

Lergotrile

This is another ergot derivative with dopaminomimetic properties; it is an effecitve antiparkinsonian agent, but hepatic toxicity occurs sufficiently frequently for the drug to have been withdrawn (Teychenne et al. 1979). This adverse reaction was not readily predictable from toxicity tests in animals – a salutory illustration of the need for careful surveillance when introducing new drugs in humans.

Lisuride

Experience with lisuride is limited, but several reports indicate that this dopaminomimetic ergot derivative has clinical properties essentially similar to those of bromocriptine (LeWitt et al. 1982b; Liebermann et al. 1980; Parkes et al. 1981). The dose range is 2.5-7.5 mg/day; although more potent than bromocriptine, its therapeutic index is similar.

Pergolide

Pergolide is another new dopaminomimetic ergot congener. Some reports claim that pergolide has a more prolonged clinical action than other dopamine agonists (Liebermann et al. 1981; Lang et al. 1982; Klawans et al. 1981), while others find it has properties very much like bromocriptine and lisuride (LeWitt et al. 1982a). Further experience is needed to resolve these somewhat controversial observations.

Cu 32-085[1]

This dopaminomimetic ergot is currently undergoing evaluation in Parkinson's disease. Initial reports are encouraging (Rinne 1983).

CQ 32-084

CQ 32-084 has been reported to have antiparkinson properties, but experience with this new ergot derivative is very limited (Rinne 1983).

Where Do Dopaminerigc Agonists Act in Parkinson's Disease?

It is generally thought that dopamine agonists achieve their antiparkinson effect in the striatum, though there is no direct evidence for this view. The striatum contains D1

---

[1] This drug has meanwhile been withdrawn from human studies (the editors).

Table 2. Summary of the actions of drugs that have been employed as dopaminergic agonists to treat Parkinson's disease

	D1 receptors	D2 receptors
Bromocriptine	-	+
Lergotrile	-	+
Lisuride	-	+
Pergolide	+	+
CU 32-085[a]	-	+
CQ 32-084	+	+

Abbreviations: + = stimulation; = inhibition.
[a] This drug has meanwhile been withdrawn from human studies (the editors).

receptors and D2 receptors. A simplified summary of the pharmacological properties of dopaminergic agonists employed to treat parkinsonism appears in Table 2. The common denominator is clearly agonism at the D2 receptor; no firm conclusion be be drawn concerning the possible role of D1 agonism.

How Do Dopaminergic Agonists Act in Parkinson's Disease?

This question will remain unanswered until more is known concerning the physiology of the dopaminergic pathways that are affected by the pathology of Parkinson's disease. One report (Stoof and Kebabian 1982) that may be relevant is the finding that in striatal tissue slices of the rat activation of D2 receptors leads to a reduction in the release of acetylcholine. This is of interest because of the well-established clinical observation that anticholinergic agents have therapeutic activity in Parkinson's disease. Another intriguing finding is antagonism between the activity of D1 and D2 receptors in the striatum. Stoof and Kebabian (1981) found, in rat striatum, that stimulation of D2 receptors leads to inhibition of the response of D1 receptors. The significance of this interaction will remain elusive until more is known about the functions of the D1 receptor.

Conclusions

Major advances have been made in characterising and classifying dopaminergic receptors, but the subject has proved particularly difficult because of the limited range of substances available as selective agonists and antagonists. Nevertheless, current evidence indicates that at least two types of dopaminergic receptors exist, and there may be more.

Artificial agonists have a useful role to play in the management of certain patients with Parkinson's disease; they appear to achieve their therapeutic effect by stimulating D2 receptors. While bromocriptine is currently the most widely employed dopaminergic agonist, a series of similar substances is undergoing clinical evaluation and initial results are encouraging. For the future, compounds with higher specificity for the different categories of dopaminergic receptor are needed, and new approaches are desirable to elucidate the challenging problems of ameliorating the psychiatric complications of late Parkinson's disease, and impeding the progress of the underlying pathology.

References

Calne DB, Teychenne PF, Claveria LE, Eastman R, Greenacre JK, Petrie A (1974) Bromocriptine in parkinsonism. Br Med J 4: 442-444

Corrodi H, Fuxe K, Hofelt T, Lidbrink P, Ungerstedt U (1973) Effect of ergot drugs on central catecholamine neurons: Evidence for a stimulation of central dopamine neurons. J Pharm Pharmacol 25: 409-412

Cotzias GC, Lawrence WH, Papavasiliou PS, Duby SE, Ginos JZ, Mena I (1972) Apomorphine and parkinsonism. Trans Am Neurol Assoc 97: 156-158

Cotzias GC, Papavasiliou PS, Tolosa ES, Mendez JS, Bell-Midura M (1976) Treatment of Parkinson's disease with aporphines. N Engl J Med 294: 567-572

Eisler T, Thorner MO, MacLeod RM, Kaiser DL, Calne DB (1981) Prolactin secretion in Parkinson disease. Neurology (NY) 31: 1356-1359

Flückiger E (1976) The pharmacology of bromocriptine. In: Bayliss RIS, Turner P, Maclay WP (eds) Pharmacological and clinical aspects of bromocriptine (Parlodel) – Proceedings of a symposium held at The Royal College of Physicians, London, 14 May 1976, MCS Consultants, Tunbridge Wells, pp 12-26

Hornykiewicz O (1975) Parkinson's disease and its chemotherapy. Biochem Pharmacol 24: 1061-1065

Hornykiewicz O (1980) Biochemical abnormalities in some extrastriatal neuronal systems in Parkinson's disease. In: Rinne UK, Klinger M, Stamm G (eds) Parkinson's disease – Current progress, problems and management, Elsevier/North-Holland, Amsterdam, pp 109-119

Kebabian JW, Calne DB (1979) Multiple receptors for dopamine. Nature 277: 93-96

Klawans HL, Tanner CM, Goetz CG, Glatt S, Nausieda P, Weiner W (1981) Pergolide mesylate therapy in Parkinson's disease: Report of a 3-month trial in 20 patients. Neurology (NY) 31/2: 133

Lang AE, Quinn N, Brincat S, Marsden CD, Parkes JD (1982) Perolide in late stage Parkinson disease. Ann Neurol 12: 243-247

Lee, T, Seeman P, Rajput A, Farley IJ, Hornykiewicz O (1978) Receptor basis for dopaminergic supersensitivity in Parkinson's disease. Nature 273: 59-61

Le Fur G, Meininger V, Baulac M, Phan T, Uzan A (1981) Recepteurs dopaminergiques lymphocytaires et maladie de Parkinson idiopathique. Rev Neurol (Paris) 137: 89-96

LeWitt PA, Ward CD, Larsen TA, Raphaelson MI, Newman RP, Foster N, Dambrosia J, Calne DB (1982a) Comparison of pergolide and bromocripte therapy in parkinsonism. Neurology (NY) 32: A181

LeWitt PA, Gopinathan G, Ward CD, Sanes JN, Dambrosia JM, Durso R, Calne DB (1982b) Lisuride versus bromocriptine treatment in Parkinson disease: A double-blind study. Neurology (NY) 32: 69-72

Lieberman A, Neophytides A, Leibowitz N, eta l. (1980) The efficacy of a potent dopamine agonist, lisuride, in Parkinson's disease (Abstr). Neurology (NY) 30: 366

Lieberman A, Goldstein M, Leibowitz M, Neophytides A, Kupersmith M, Pact V, Kleinberg D (1981) Treatment of advanced Parkinson disease with pergolide. Neurology (NY) 31: 675-682

Parkes JD, Schachter N, Marsden CD, Smith B, Wilson A (1981) Lisuride in parkinsonism. Ann Neurol 9: 48-52

Quik M, Spokes EG, Mackay AVP, Bannister R (1979) Alteration in (^{3}H) spiperone binding in human caudate nucleus, substantia nigra and frontal cortex in the Shy-Drager syndrome and Parkinson's disease. J Neurol Sci 43: 429-437

Reisine TD, Fields JZ, Yamamura HI, Bird ED, Spokes E, Schreiner PS, Enna SJ (1977) Neurotransmitter receptor alterations in Parkinson's disease. Life Sci 21: 335-344

Rinne UK, Lonnberg P, Koskinen V (1981) Dopamine receptors in parkinsonian brains. J Neural Transm 51: 97-109

Rinne UK (1983) New ergot derivatives in the treatment of Parkinson's disease. In: Calne DB, McDonald RJ, Horowski R, Wuttke W (eds) Lisuride and other dopamine agonists. Raven, New York

Schwab RS, Amador LV, Lettvin JY (1951) Apomorphine in Parkinson's disease. Trans Am Neurol Assoc 76: 251-253

Seeman P (1980) Brain dopamine receptors. Pharmacol Rev 32: 229-313

Setler PE, Sarau HM, Zerkle CL, Saunders HL (1978) The central effects of a novel dopamine agonist. Eur J Pharmacol 50: 419-430

Sokoloff P, Matres MD, Schwartz JP (1980) Three classes of dopamine receptor (D-2, D-3, D-4) identified by binding studies with ^{3}H-apomorphine and ^{3}H-domperidone. Naunyn Schmiedebergs Arch Pharmacol 315: 89-102

Spano PF, Govoni S, Trabucchi M (1978) Studies on the pharmacological properties of dopamine receptors in various areas of the central nervous system. In: Roberts PF, Woodruff GN, Iversen LL (eds) Advances in biochemical psychopharmacology, vol 19. Raven, New York, pp 155-165

Stoof JD, Kebabian JW (1981) Opposing roles for D-1 and D-2 dopamine receptors in efflux of cyclinc AMP from rat neostriatum. Nature 294: 366-368

Stoof, JD, Kebabian JW (1982) Independent in vitro regulation of D-2 dopamine receptor of dopamine-stimulated efflux of cyclic AMP and K^+ stimulated release of acetylcholine from rat neostriatum. Brain Res 250: 263-270

Teychenne PF, Jones EA, Ishak KG, Calne DB (1979) Hepatocellular injury with distinctive mitochondrial changes induced by lergotrile mesylate: A dopaminergic ergot derivative. Gastroenterology 76: 575-583

Tsuruta K, Frey EA, Grewe CW, Cote TE, Eskay RL, Kebabian JW (1981) Evidence that LY-141865 specifically stimulates the D-2 dopamine receptor. Nature 292: 463-465

Winkler MH, Berl S, Whetsell WO, Yahr MD (1980) Spiroperidol binding in the human caudate nuclear. J Neural Transm (Suppl) 16: 45-51

Brain Dopaminergic Receptor Changes in Parkinson's Disease in Relation to the Disease Process and Treatment

U.K. Rinne, J.O. Rinne, J.K. Rinne and K. Laakso

Introduction

A progressive loss of substantia nigra dopaminergic neurons and a deficiency of dopamine in the striatum are the most characteristic and prominent changes in the pathophysiology of Parkinson's disease (Ehringer and Hornykiewicz 1960). Besides these presynaptic changes of dopaminergic neurons, in recent years there have also been found alterations in dopaminergic receptors in the striatum (Reisine et al. 1977; Lee et al. 1978a; Rinne et al. 1979, 1981; Winkler et al. 1980) and in the substantia nigra (Quik et al. 1979) of patients with Parkinson's disease. The present paper is concerned with changes in brain dopaminergic receptors in Parkinson's disease in relation to the disease process and treatment.

Striatal Dopaminergic Receptors

In the author's studies (Rinne et al. 1979, 1981) the dopaminergic receptor sites were determined by the specific binding of spiperone ^{3}H. Thus in the striatum, it was mainly D2 receptors according to the classification of Seeman (1980) which were identified. With regard to the behaviour of striatal binding of spiperone ^{3}H, it was possible to divide parkinsonian patients into two subgroups, since either an increase or a decrease in the number of D2 receptors was found. Furthermore, changes in the dopaminergic receptors in the parkinsonian nucleus accumbens were similar to those in the striatum (Rinne 1982).

In contrast to most untreated patients, there was found a normal or decreased number of striatal D2 receptors in parkinsonian patients treated with levodopa, using the binding of spiperone ^{3}H (Rinne et al. 1979, 1981) or haloperidol ^{3}H (Lee et al. 1978a). This suggests that levodopa treatment can reverse dopaminergic supersensitivity in the striatum. On the other hand, treatment with neuroleptic drugs caused a great increase in the number of striatal D2 receptors both in the parkinsonian patients and in the controls. Similar changes after neuroleptic treatment have been demonstrated in schizophrenic patients (Lee et al. 1978b; Owen et al. 1978) and in animals (Burt et al. 1977; List and Seeman 1979).

As in post-mortem brain studies, similar changes in D2 receptor function have also been found in our in vivo studies when responses of prolactin secretion to TRH stimulation after suppression induced with bromocriptine were used as an index of dopaminergic receptor function. After the administration of bromocriptine, TRH injection

Table 1. Binding (fmol per milligram protein) of apomorphine ^3H (D3 binding sites) in parkinsonian and control striatum. Patients without levodopa treatment. Mean ± standard error

Group	Caudate nucleus	Putamen
Controls	53 ± 5 (38)	48 ± 5 (33)
Parkinsonian patients		
Without neuroleptics	32 ± 8* (14)	45 ± 5 (11)
With neuroleptics	59 ± 22 (5)	62 ± 16 (7)

* $P < 0.05$ as compared with control.

induced a significantly higher prolactin response in parkinsonian patients suffering from daily fluctuations in performance than in other parkinsonian patients or in the controls. Furthermore, in parkinsonian patients without levodopa treatment the prolactin response was significantly lower than in the controls (Laihinen et al. 1983).

Besides changes in D2 receptors, there is recent evidence on the reduction of D3 binding sites in the parkinsonian striatum by using dopamine ^3H binding (Lee et al. 1981). Similar findings have been obtained in the author's recent study (Tables 1 and 2) using apomorphine ^3H binding under the conditions specified by Lee et al. (1978a). This suggests that at least some of the D3 binding sites are located on the presynaptic terminals of nigral neurons. Furthermore, in contrast to D2 receptors, the D3 sites in the parkinsonian striatum seem to be less sensitive to treatment with neuroleptic drugs.

Relationships Between Striatal Dopaminergic Receptors and Parkinsonian Clinical Variables

Special attention was paid to relating dopaminergic receptor changes to the clinical variables of the patients. Most of the patients were severely disabled during the final period of life and therefore it is not possible to obtain absolute correlations (Table 3). However, it was evident that the disability of the patients, dementia, and the loss of beneficial response to levodopa, were most closely related to a decrease in striatal D2 receptors. On the other hand, dyskinesias, daily fluctuations in performance, and psychotic episodes, together with neuroleptic medication, were especially associated with an increase in the number of striatal D2 receptors (Rinne et al. 1979, 1981).

These findings suggest that besides the progressive degeneration of presynaptic nigral dopaminergic neurons, a loss of postsynaptic dopaminergic receptors in the striatum may also contribute, at least partly, to the deteriorating response of parkinsonian patients to levodopa long-term treatment. Furthermore, it is obvious that changes in the response of striatal dopaminergic receptors may be one of the factors contributing

Table 2. Binding (fmol per milligram protein) of apomorphine ^3H (D3 binding sites) in parkinsonian and control striatum. Patients with chronic levodopa treatment. Mean ± standard error

Group	Caudate nucleus	Putamen
Controls	53 ± 5 (38)	48 ± 5 (33)
Parkinsonian patients		
Without neuroleptics	24 ± 4** (6)	27 ± 9* (4)
With neuroleptics	42 ± 10 (12)	46 ± 4 (12)

* $P < 0.05$ as compared with control. ** $P < 0.001$ as compared with control.

to the mechanisms of daily fluctuations in performance occurring during chronic levodopa treatment. However, comparative studies of dopamine concentration and dopaminergic receptors have shown that in the parkinsonian striatum there is a greater loss of presynaptic dopaminergic neurons than postsynaptic dopaminergic receptor sites (Lee et al. 1978a; Rinne et al. 1979). Thus in advanced parkinsonian patients with a deteriorating response to levodopa there still seem to be enough dopaminergic receptors in the striatum for drugs directly stimulating the dopaminergic receptors to alleviate the parkinsonian disability. Indeed, many recent clinical studies have shown that this really is the case (see Rinne 1978, 1980, 1981).

Interaction with Neurotransmitter Receptors and Therapeutic Responses of Dopaminergic Agonists in Parkinson's Disease

The therapeutic responses of dopaminergic agonists in parkinsonian patients seem to depend on drug interaction primarily with dopaminergic receptors. Indeed, ergot derivatives which have been investigated in the treatment of Parkinson's disease, such as bromocriptine, CM 29-712, CQ 32-084, CU 32-085[1], lisuride and pergolide, have dopaminergic properties which have been demonstrated in a number of in vivio and in vitro tests. In certain respects they differ in their specificity for the different subtypes of dopaminergic receptors, and in addition they interact with other neurotransmitter receptors (Pieri et al. 1978; Horowski and Wachtel 1979; Azuma and Oshino 1980; Wong and Reid 1980; Wong et al. 1980; Lemberger et al. 1980; Kehr, personal communication 1981; Markstein 1981, personal communication, 1981; Ringwald et al. 1082). Bromocriptine, CU 32-085[1] and lisuride stimulate the D2 receptors and are antagonistic

1 This drug has meanwhile been withdrawn from human studies (the editors).

Table 3. Relationship between clinical variables of parkinsonian patients and changes in striatal dopaminergic receptors

Variable	Striatal dopaminergic receptors	
	Decreased	Increased
1. Disability	++	+
2. Degree of dementia	++	+
3. Levodopa – loss of response	++	+
4. Dyskinesias	+	++
5. On-off phenomena	+	++
6. Psychotic episodes with neuroleptics	0	++

Table 4. Neurotransmitter receptor profile of various dopaminergic agonists. Agonistic (+) or antagonistic (-) effect. Prolactin inhibitory potency (PRL) as compared with bromocriptine

Compound	0-1	0-2	5-HT-1	$\alpha 1 + \alpha 2$	PRL
Bromocriptine	-	+	-	-	1
CU 32-085[b]	-	+	+	-	5
Lisuride	+/-?	+	+/-[a]	-	12
CQ 32-084	+	+	+	-	10
Pergolide	+	+	-	-	20
CM 29-712	+	-	+	-	< 0.1

[a] + Centrally / - peripherally.
[b] This drug has meanwhile been withdrawn from human studies (the editors).

to D1 receptors. CQ 32-084 and pergolide appear to stimulate both D1 and D2 receptors, whereas CM 29-712 mainly stimulates D1 receptors. In addition, bromocriptine blocks and the other ergot derivatives stimulate serotonergic receptors linked to an adenylate cyclase. All these compounds have antagonistic effects on α 1- and α 2-adrenoceptors. Moreover, there are differences in the prolactin inhibitory potencies of these ergot derivatives as compared with bromocriptine (Table 4).

Although the ergot derivatives described have distinctive effects on multiple neurotransmitter receptors, it is evident that in Parkinson's disease the therapeutic response is elicited primarily by stimulation of D2 receptors (Schachter et al. 1981). The author's studies (Rinne 1980, 1983a, b) have shown that the therapeutic profile of ergot derivatives seems to be similar in the primary treatment of Parkinson's disease. The improvement in tremor was significantly better than that in rigidity and hypokinesia. The degree of therapeutic efficacy of the dopaminergic agonists studied was roughly similar, although the compounds with both D2 and D1 stimulating properties seem to have somewhat greater antiparkinsonian effect (Table 5). However, without double-blind comparison it is not possible to draw valid conclusions.

Table 5. Therapeutic responses to various dopaminergic agonists. Relativ effect during 3 months treatment

Drug	Number of patients	Dose (mg)	Therapeutic response				
			Total disability	Tremor	Rigidity	Hypo-kinesia	Mood elevation
Bromocriptine	72	27.3±2.5 (3.75-120.0)	+	++	+	+	+
CU 32-085[a]	11	18.5±4.9 (1.5-40.0)	+	++	+	+	+++
Lisuride	10	1.5±0.2 (0.6-2.4)	+	++	+	+	+
CQ 32-084	8	29.3±8.9 (3.0-50.0)	++	+++	+	++	+
Pergolide	20	2.9±0.6 (0.3-6.0)	+	+++	+	+	+
CM 29-712	8	23.0±6.9 (7.5-50.0)	+	++	+	+	-

[a] This drug has meanwhile been withdrawn from human studies (the editors).

Table 6. Improvement (%) of parkinsonian patients during 3 months treatment with levodopa and benserazide, bromocriptine, lisuride or pergolide

Variable	Treatment			
	Levodopa + benserazide (N = 65)	Bromocriptine (N = 72)	Lisuride (N = 9)	Pergolide (N = 20)
Total disability	38 ± 2	18 ± 2***	22 ± 6*	21 ± 3***
Tremor	44 ± 3	49 ± 4	56 ± 9	60 ± 5**
Rigidity	37 ± 3	9 ± 2***	13 ± 6**	9 ± 2***
Hypokinesia	38 ± 3	16 ± 4***	21 ± 5**	22 ± 4**

* $P < 0.05$ as compared with levodopa + benserazide.
** $P < 0.01$ as compared with levodopa + benserazide.
*** $P < 0.001$ as compared with levodopa + benserazide.

Table 7. Therapeutic responses to various dopaminergic agonists during 3 months treatment in parkinsonian patients with a deteriorating response and daily fluctuations in performance

Drug	Number of patients	Dose (mg)	Therapeutic responses	
			Parkinsonian disability	On-off phenomena
Bromocriptine	41	33.0 ± 4.3 (10-160)	+	+
CU 32-085[a]	10	9.5 ± 1.6 (3-15)	+	++
Lisuride	10	1.5 ± 0.4 (0.6-4.0)	+	++
CQ 32-084	7	3.4 ± 1.2 (0.5-10.0)	+	++
Pergolide	18	0.5 ± 0.1 (0.1-1.5)	+	++
CM 29-712	8	18.9 ± 4.3 (1.0-30.0)	+	+

[a] This drug has meanwhile been withdrawn from human studies (the editors).

According to the author's findings the therapeutic efficacy of dopaminergic agonists on parkinsonian disability in previously untreated patients seems to be significantly less than that of levodopa. However, the mean degree of improvement in tremor was equal with both these treatment regimens (Table 6). Thus levodopa is still the most effective antiparkinsonian agent as a primary treatment of Parkinson's disease.

Although dopaminergic agonists as a monotherapy seem to have only a limited value it has been clearly shown by us (Rinne et al. 1975, 1977; Rinne and Marttila 1978; Rinne 1980, 1983a, b) and by others (Calne et al. 1974, 1978; Lieberman et al. 1976, 1980; Parkes et al. 1976, 1981, Grön 1977; Jansen 1978; Ludin et al. 1978, Fahn et al. 1979; Gopinathan et al. 1981) that they are significant and valuable adjuvant therapy to levodopa in patients with daily fluctuations in performance. All the dopaminergic agonists studied showed roughly similar beneficial effects in the treatment of daily fluctuations in performance, although CM 29-712 and bromocriptine seem to be somewhat less effective than the other new compounds (Table 7). However, without double-blind comparisons it is not possible to draw valid conclusions.

Conclusions

1. In Parkinson's disease, not only is there a degeneration of nigrostriatal dopaminergic neurons, but significant changes also occur in striatal dopaminergic receptors.

2. There is evidence of a loss of striatal D3 receptors in Parkinson's disease.

3. There seem to be at least two subgroups of parkinsonian patients with regard to the behaviour of striatal D2 receptors: (a) those with a loss of D2 receptors; and (b) those with an increase in the number of D2 receptors.

4. Clinically, the patients with a low number of striatal D2 receptors were more disabled and had lost the beneficial response to levodopa. In contrast, dyskinesias, daily fluctuations in performance and neuroleptic medication were especially related to the supersensitivity of striatal D2 receptors.

5. Increased responses of prolactin secretion to TRH in vivo also suggested decreased D2 receptor function in parkinsonian patients with recent onset and supersensitivity in patients with daily fluctuations in performance.

6. In advanced parkinsonian patients with a deteriorating response to levodopa there seem still to be enough dopaminergic receptors in the striatum for drugs directly stimulating the dopaminergic receptors to alleviate the parkinsonian disability and the daily fluctuations in performance.

7. There were no definite and clear-cut relations between therapeutic profiles of dopaminergic agonists and effects on dopaminergic receptors. The therapeutic responses of dopaminergic agonists were elicited by compounds primarily stimulating the D2 or D1 receptors. However, it is evident that drugs with both D2 and D1 stimulating properties have more antiparkinsonian effect.

8. The therapeutic response of previously untreated parkinsonian patients to the dopaminergic agonists seems to be less than that to levodopa. However, the degree of improvement in tremor was equal with both treatment regimens. Thus, levodopa is still the most effective antiparkinsonian agent as a primary treatment of Parkinson's disease.

9. At the present the main indications for dopaminergic agonists seem to be a combined treatment with levodopa in patients with a deteriorating response to levodopa and daily fluctuations in performance.

Acknowledgment. This study was supported by a grant from the Sigrid Jusélius Foundation, Finland.

References

Azuma H, Oshino N (1980) Stimulatory action of lisuride on dopamine sensitive adenylate cyclase in the rat striatal homogenate. Jpn J Pharmacol 30: 629-639

Burt DR, Creese I, Snyder SH (1977) Antischizophrenic drugs: Chronic treatment elevates dopamine receptor binding in brain. Science 196: 326-328

Calne DB, Teychenne PF, Claveria LE, Eastman, R, Greenacre JK, Petrie A (1974) Bromocriptine in parkinsonism. Br Med J 4: 442-444

Calne DB, Plotkin C, Williams AC, Nutt JG, Neophytides A, Teychenne PF (1978) Long-term treatment of parkinsonism with bromocriptine. Lancet 1: 735-738

Ehringer H, Hornykiewicz O (1960) Verteilung von Noradrenalin und Dopamine (3-hydroxytyramin) im Gehirn des Menschen und ihr Verhalten bei Erkrankungen des extrapyramidalen Systems. Klin Wochenschr 38: 1236-1239

Fahn S, Cote LJ, Snider SR, Barrett RE, Isgreen WP (1970) The role of bromocriptine in the treatment of parkinsonism. Neurology (NY) 29: 1077-1083

Gopinathan G, Teräväinen H, Dambrosia JM, Ward CD, Sanes JN, Stuart WK, Evarts EV, Calne DB (1981) Lisuride in parkinsonism. Neurology (NY) 31: 371-376

Grön U (1977) Bromocriptine versus placebo in levodopa treated patients with Parkinson's disease. Acta Neurol Scand 56: 269-273

Horowski R, Wachtel H (1979) Pharmacological effects of lisuride in rodents mediated by dopaminergic receptors: mechanism of action and influence of chronic treatment with lisuride. In: Fuxe K, Calne DB (eds) Dopaminergic ergot derivatives and motor function. Pergamon, Oxford, pp 237-251

Jansen ENH (1978) Bromocryptine in levodopa response-losing parkinsonism. Eur Neurol 17: 92-99

Laihinen A, Irjala K, Rinne UK (1983) Prolactin response as an index of dopamine receptor function in Parkinson's disease. In: Endroczi E et al. (eds) Neuropeptides and psychosomatic processes. Akademiai Kiado, Budapest, pp 653-659

Lee T, Seeman P, Rajput A, Farley IJ, Hornykiewicz O (1978a) Receptor basis for dopaminergic supersensitivity in Parkinson's disease. Nature 273: 59-61

Lee T, Seeman P, Tourtellotte WW, Farley IJ, Hornykiewicz O (1978b) Bindung of ^{3}H-neuroleptics and ^{3}H-apomorphine in schizophrenic brains. Nature 274: 897-900

Lee T, Seeman P, Hornykiewicz O, Bilbao J, Deck J, Tourtellotte W (1981) Parkinson's disease: low density and presynaptic location of D_3 dopamine receptors. Brain Res 212: 494-498

Lemberger L, Crabtree R, Callaghan JT (1980) Pergolide, a potent long-acting dopamine-receptor agonist. Clin Pharmacol Ther 27: 642-651

Lieberman A, Zolfaghari M, Boal D, Hassouri H, Vogel B, Battista A, Fuxe K, Goldstein M (1976) The antiparkinsonian efficacy of bromocriptine. Neurology (Minneap) 26: 405-409

Lieberman A, Kupersmith M, Neophytides A, Casson I, Durso R, Hoo Foo S, Khayali M, Bear G, Goldstein M (1980) Long-term efficacy of bromocriptine in Parkinson disease. Neurology (NY) 30: 518-523

List SJ, Seeman P (1979) Dopamine agonists reverse the elevated ^{3}H-neuroleptic binding in neuroleptic-pretreated rats. Life Sci 24: 1447-1452

Ludin HP, Ringwald E, Lorincz P (1978) Bromocriptine in Parkinson's disease. Lancet 2: 578-579

Markstein R (1981) Neurochemical effects of some ergot derivatives: A basis for their antiparkinsonian actions. J Neural Transm 51: 39-59

Owen F, Cross AJ, Grow TJ, Longden A, Poulter M, Riley GJ (1978) Increased dopamine-receptor sensitivity in schizophrenia. Lancet 1: 223-226

Parkes JD, Marsden CD, Donaldson I (1976) Bromocriptine treatment in Parkinson's disease. J Neurol Neurosurg Psychiatry 39: 184-193

Parkes JD, Schachter M, Marsden CD, Smith B, Wilson A (1981) Lisuride in parkinsonism. Ann Neurol 9: 48-52

Pieri L, Keller HH, Burkhard W, de Prada M (1978) Effects of lisuride and LSD on cerebral monoamine systems and hallucinosis. Nature 272: 278-280

Quik M, Spokes EG, Mackay AVP, Bannister R (1979) Alterations in (^{3}H)spiperone binding in human caudate nucleus, substantia nigra and frontal cortex in the Shy-Drager syndrome and Parkinson's disease. J Neurol Sci 43: 429-437

Reisine TD, Fields JZ, Yamamura HI, Bird ED, Spokes E, Schreiner PS, Enna SJ (1977) Neurotransmitter receptor alterations in Parkinson's disease. Life Sci 21: 335-344

Ringwald E, Hirt D, Markstein R, Vigouret JM (1982) Dopaminrezeptoren-Stimulatoren in der Behandlung der Parkinsonkrankheit. Nervenarzt 53: 67-71

Rinne UK (1978) Recent advances in research on Parkinsonism. Acta Neurol Scand 57: 77-113

Rinne UK (1980) Recent advances in the treatment of Parkinson's disease. Acta Neurol Scand (Suppl 78) 62: 103-121

Rinne UK (1981) Treatment of Parkinson's disease: Problems with a progressing disease. J Neural Transm 51: 161-174

Rinne UK (1982) Brain dopamine receptors in Parkinson's disease: Involvement with clinical features and therapeutic responses. In: Riederer P, Usdin E (eds) Human brain tissue. Macmillan, London, pp 303-317

Rinne UK (1983a) Dopamine agonists in the treatment of Parkinson's disease. Adv Neurol 37: 141-150

Rinne UK (1983b) New ergot derivatives in the treatment of Parkinson's disease. In: Calne DB, Horowski R, McDonald TR, Wuttke W (eds) Lisuride and other dopamine agonists. Raven, New York, pp 431-442

Rinne UK, Marttila RJ (1978) Brain dopamine receptor stimulation and the relief of Parkinsonism. Relationship between bromocriptine and levodopa. Ann Neurol 4: 263-267
Rinne UK, Sonninen V, Marttila R (1975) Dopaminergic agonist effects on parkinsonian clinical features and brain monoamine metabolism. Adv Neurol 9: 383-392
Rinne UK, Marttila RJ, Sonninen V (1977) Brain dopamine turnover and the relief of Parkinsonism. Arch Neurol 34: 626-629
Rinne UK, Sonninen V, Laaksonen H (1979) Responses of brain neurochemistry to levodopa treatment in Parkinson's disease. Adv Neurol 24: 259-274
Rinne UK, Lönnberg P, Koskinen V (1981) Dopamine receptors in the parkinsonian brain. J Neural Transm 51: 97-106
Schachter M, Bédard R, DeBono AG, Jenner P, Marsden CD, Price P, Parkes JD, Keenan J, Smith B, Rosenthaler J, Horowski R, Dorrow R (1981) The role of D-1 and D-2 receptors. Nature 286: 157-159
Seeman P (1980) Brain dopamine receptors. Pharmacol Rev 32: 230-313
Winkler MH, Berl S, Whetsell WO, Yahr MD (1980) Spiroperidol binding in the human caudate nucleus. J Neural Transm 16: 45-51
Wong DT, Reid LR (1980) Activation of adenylaste cyclase in rat striatum by an ergoline dopamine agonist, pergolide. Commun Psychopharmacol 4: 269-275
Wong DT, Bymaster FP, Lane PT, Kau D, Bach NJ, Kornfeld EC (1980) Binding of (^{3}H)pergolide mesylate to dopamine receptors of mammalian brains. Res Commun Chem Pathol Pharmacol 30: 195-210

Analysis of Akinesia

H. Narabayashi

Introduction

Akinesia is one of the most frequently used terms in the symptomatology of parkinsonism. The term includes various symptoms, and sometimes it is not easy to define akinesia clearly and strictly, especially when its generating mechanism is discussed (Barbeau 1971; Hornykiewicz 1972). In this paper I would like to propose at least three different subgroups of akinesia in the clinical sense, considering their physiological and pharmacological mechanisms (Narabayashi 1980). Parkinsonian facies, poor blinking, low voice and dysarthria, stuttering festination of speech, lack of arm-swing in walking, poverty and slowness of movement, awkwardness in coordinated movements and many other difficulties in motor performance of daily activities are included in the term akinesia. Not all of these symptoms can be considered to be due to one single mechanism, but may be due to several different mechanisms in their generation. Group A consists of symptoms which belong to the category of slowness and awkwardness of movement, especially of coordinated movement, and group B is the group of symptoms belonging to the category of poverty or lack of movement. Group C is the so-called freezing phenomenon at the start of movement or festination of repetitive movements.

Subgroups of Akinesia

Group A akinesia is interpreted as being secondary to the existence of marked rigidity of muscles (Ohye et al. 1965). Most simply it is recognized in the difficulty and slowness of repetitive flexion and extension movement of the highly rigid forearm at the elbow joint. Electromyographically, when voluntary contraction of the biceps brachii muscle is tried in such a patient, it is always accompanied by cocontraction of the antagonistic muscle, the triceps brachii, Westphal's phenomenon. Because of such tonic involuntary antagonist activity, the effect of voluntary contraction of the biceps muscle cannot be as strong and quick as in normal subjects in whom inhibition of antagonist activity produces the phenomenon known as reciprocal inhibition. This type of difficulty may be considered as due to disturbance of reciprocal innervation and a similar observation can also easily be made when passive stretching of one muscle, agonist or antagonist, is tried. Such difficulty can quickly be returned to normal by reducing or abolishing rigidity of muscles by pharmacological or neurosurgical means. This change is particularly impressive when marked rigidity is reduced by ventrolateral thalamotomy to the normal tone level. Within a few seconds of alleviation of rigidity, the patient almost always becomes capable of repetitive flexion-extension movement or other more complex coordinated movement just as freely and normally on the operating table. Electromyographically, previously observed discharge of the antagonist muscle disappears, facilitating normal and selective voluntary contraction of the agonist without interference of the antagonist. From these observations, it is suggested that in this type of akinesia, group A, the slowness and awkwardness of delicate voluntary movement is caused by and secondary to the existence of muscle rigidity.

Group B akinesia is a well-known type of akinesia, i.e. poverty or lack of movement. Martin (1967) has beautifully described this characteristic feature in parkinsonian patients. Patients seem to have reduced motivation or drive of movement, though they can perform quite strong muscle contraction and movement, albeit slowly and unskilfully, when requested.

For instance, when the patient is sitting waiting for a friend, unintended small movements such as swinging the upper leg when the legs are crossed, or movements of hands and fingers, which are commonly seen in normal subjects, are lacking. When listening to music, the patient has very little tendency to move or swing hands, arms or legs in rhythm. The intracerebral neuronal mechanism responsible for these changes in parkinsonian patients are not known precisely, but it is suggested that such poverty of movement is due to dopamine deficiency. It is also established that this type of akinesia is different from that of group A akinesia secondary to rigidity, and has no relation to the existence of rigidity. Cases in which rigidity has been markedly improved by stereotactic surgery often still show this group B akinesia, indicating that it is quite independent of the existence of rigidity. The most important benefit of L-dopa therapy is that the drug can modify and improve group B akinesia, which cannot be changed by surgical methods. It also provides the reason that L-dopa therapy is considered today to have a wider spectrum of effect than surgery and is therefore treatment of choice for parkinsonian patients.

Group C akinesia is the symptom designated by the term "freezing". Freezing is most often observed in gait, but it also appears in speech and handwriting. For example, the patient cannot begin walking forward or halts in the act of walking after festination. This difficulty becomes greater, when the patient attempts to go through a narrow gate or door. However, the patient can easily step over objects, just as the normal subject does. There is no difficulty in going up and down stairs. Therefore, this peculiar difficulty in gait is named *kinesie paradoxale.* Such freezing is often observed in cases with slighter rigidity rather than in those with marked rigidity and even appears without any sign of rigidity or tremor, as described in the report on "cases of pure akinesia" by Narabayashi et al. (1976). This peculiar symptom seems to have no relation to the existence of rigidity or of tremor. Cases of pure akinesia are without any sign of muscular hypertonia and are often even hypotonic. For analysis of *kinesie paradoxale,* Nakamura et al. (1976) applied the finger tapping test. These patients cannot perform finger tapping at frequencies over 5 Hz at maximum effort, though 7-8 Hz tapping is easy for the age-matched control subjects. A very specific feature in the tapping test in response to given signals (sound or flash) is that these patients can follow exactly the frequency of the signal if it is below 2 Hz but cannot follow signals over 2.5 Hz at all. In the latter instance, tapping with a frequency higher than 2.5 Hz is possible only around 5 Hz. There seems to exist a very specific neuronal mechanism in the CNS of these patients with freezing and this may enable the patient to make rhythmic repetitive movement only around 5 Hz (which is similar to tremor frequency) or below 2 Hz. This tendency of freezing has been observed to become more prominent and more severe in parkinsonian patients of longer disease duration than in the relatively earlier stages of the disease. And even when other symptoms of parkinsonism have well been controlled by long-term L-dopa and other agents, sometimes with stereotactic surgery, the symptom of freezing gradually becomes manifest and worsens, seriously affecting the patient's daily life. Such freezing cannot be modified or improved by L-dopa, and it is often worsened by further increasing the dosage of L-dopa. It is therefore suggested that freezing is not due to dopamine deficiency.

Therapy of Freezing

Nagatsu et al. (1979, 1982) and Narabayashi et al. (1981) have demonstrated that the activity of dopamine-β-hydroxylase (DBH) in CSF is much more severely lowered in cases with long disease duration (more than 5 years duration), which may indicate that biosynthesis of norepinephrine in the CNS is significantly decreased in the later stages of the illness. Based on the working hypothesis that the freezing phenomenon in the later stages of parkinsonism, which does not respond to L-dopa therapy, may be due to deficiency of norepinephrine because of the severely lowered DBH activity, we have started to administer DL- or L-*threo*-3,4-dihydroxyphenylserine (DL- or L-*threo*-DOPS), the precursor of norepinephrine, to these patients, aiming to compensate the deficient norepinephrine (Narabayashi et al. 1981, 1983). When DL-*threo*-DOPS is used, biological activity is estimated to be about half of that of L-*threo*-DOPS. Starting medication with DL-*threo*-DOPS 600-800 mg/day together with 60-90 mg carbidopa as peripheral decarboxylase inhibitor, marked improvement of freezing symptoms is observed from the second day, as reported in detail by Narabayashi et al. (1981, 1983). In about 80% of the 12 reported cases with freezing symptoms in gait, writing or speech, these symptoms were alleviated or markedly decreased. This improvement by DL- or L-*threo*-DOPS, enabling the patient to walk normally, is quite impressive and comparable to the effect of L-dopa on groups A and B akinesia. From these observations, we would tentatively suggest that freezing symptoms in the later stages of parkinsonism are due to norepinephrine deficiency in the CNS.

Conclusions

Akinesia in parkinsonism, which is a relatively vague term, is analysed into three subgroups because of their pharmacological and physiological characteristics. Dopamine is important for groups A and B and norepinephrine is suggested to be of importance for generation of group C akinesia. Parkinson's disease, if it is observed over a long period, e.g. 10-15 years, is revealed as a disorder of catecholamine metabolism (dopamine and norepinephrine). In the earlier stages, the dopaminergic system may be more seriously affected and in the later stages, progressive disturbance of the noradrenergic system will be added. The existence of well-known neuropathological findings in the pigmented neurons of the substantia nigra and of the locus coeruleus may indicate the importance of these two metabolites.

In the daily practice of clinical neurology with parkinsonian patients, the impression gradually increases that the effects of psychic or psychological changes on motor performance seem to be equally important as changes in motor function itself. Even *kinesie paradoxale* may to some extent be understood on the basis of psychophysiology as "apraxia of gait". Furthermore, poor, slow and akinetic features of psychomotor function be interpreted as a fourth type of akinesia. However, the author has no idea how to analyse or evaluate this type of difficulty objectively; it remains an important topic for future research.

References

Barbeau A (1971) Long-term side-effects of Levodopa. Lancet 1: 395

Hornykiewicz O (1972) Biochemical and pharmacological aspects of akinesia. In: Siegfried J (ed) Parkinson's disease, vol 1. Huber, Bern, pp 127-149

Martin J (1967) The basal ganglia and posture. Pitman, London, p 17

Nagatsu T, Kato T, Nagatsu I, Kondo Y, Inagaki S, Iizuka R, Narabayashi H (1979) Catecholamine-related enzymes in the brain of patients with parkinsonism and Wilson's disease. Adv Neurol 24: 283-292

Nagatsu T, Wakui Y, Kato T, Fujita K, Kondo T, Yokochi F, Narabayashi H (1982) Dopamine beta-hydroxylase activity in cerebrospinal fluid of parkinsonian patients. Biomed Res 3: 95-98

Nakamura R, Nagasaki H, Narabayashi H (1976) Arrhythmokinesia in parkinsonism. In: Birkmayer W, Hornykiewicz O (eds) Advances in parkinsonism. Hoffmann-La Roche, Basel, pp 258-268

Narabayashi H (1980) Clinical analysis of akinesia. J Neural Transm (Suppl) 16: 129-136

Narabayashi H, Nakamura R (1981) The freezing phenomenon – Problems in long-term L-dopa treatment for parkinsonism. In: Rose FC, Capildeo R (eds) Research progress in parkinson's disease. Pitman, Tunbridge Wells, pp 248-253

Narabayashi H, Imai H, Yokochi M, Hirayama K, Nakamura R (1976) Cases of pure akinesia without rigidity and tremor and with no effect by L-dopa therapy. In: Birkmayer W, Hornykiewicz O (eds) Advances in parkinsonism. Hoffmann-La Roche, Basel, pp 335-342

Narabayashi H, Kondo T, Hayashi A, Suzuki T, Nagatsu T (1981) L-threo-3,4-dihydroxyphenylserine treatment for akinesia and freezing of parkinsonism. Proc Jpn Acad (B) 57: 351-354

Narabayashi H, Kondo T, Hayashi A, Suzuki T, Nagatsu T (1983) L-threo-3,4-dihydroxyphenylserine for treatment of freezing and akinesia of parkinsonism. In: Yahr MD (ed) Current concepts of Parkinson disease and related disorders. Excerpta Medica, Princeton

Ohye C, Tsukahara H, Narabayashi H (1965) Rigidity and disturbance of reciprocal innervation. Confin Neurol 26: 24-40

Observations on Voluntarily Induced Automatic Motor Behaviour in Parkinsonism

J.P.W.F. Lakke and W. van den Burg

Introduction

Turning over in bed, creeping, righting and normal locomotion are voluntarily induced complex automatic movements, commonly considered as part of postural mechanisms. There are reasons to question this opinion. Parkinsonian patients, adequately treated with L-dopa compounds and with an acceptable therapeutic response with respect to tremor, rigidity and akinesia, often report no improvement, or, as time passes, even deterioration of one or more of these voluntarily induced automatic movements. A trial to improve kneeling, rotation in the horizontal plane and righting by intensive rehabilitation training failed. Other patients report fragmentation strategies for certain automatic movements. Apparently at least part of voluntarily induced automatic motor behaviour is not dependent upon the dopaminergic system and this raises the question whether the neuroanatomical substrate for these automatic movements is to be found in the basal ganglia.

Parkinson's disease comprises a variety of symptoms, occurring in various combinations and with different severity. Besides tremor, rigidity and akinesia, vegetative disturbances, speech and writing disorders, posture, balance and mobility can become severely handicapped. Pharmacotherapy affects the various symptoms differently, stereotactic surgery aims particularly at the alleviation of the tremor. It is an intriguing problem why physiotherapy in parkinsonian patients produces such poor results. Since the advent of L-dopa, with consequently presumed restoration of dopamine deficiencies in the diseased nigrostriatal pathways, physiotherapeutic efforts have generally failed to improve mobility. Although the patients themselves often mention benefit from the physiotherapy this is not reflected in the results of assessment of motor abilities with ADL (activities of daily living) scales (Franklyn et al. 1981). When performing everyday activities, this requires normal postural and locomotor function. The pharmacotherapeutically well-treated Parkinson patients are usually able to perform such skilled manual motor tasks as instrument manipulation, piano playing (albeit at slower tempo), but may be deficient in performing more automatically executed movements like walking, turning, kneeling and rising. Such stereotyped motor patterns, can be the subject of comparative study, because in the western european culture these motor abilities are acquired in early life, independent of social status and are executed frequently.

Patients and Methods

With regard to this automatic motor behaviour we studied 41 ambulant parkinsonian patients. The patients were attending an outpatient clinic for movement disorders and were included in the study if they admitted difficulties with turning over in bed, when asked. The group comprises 23 males, 18 females, mean age 66.6 years, standard deviation 11.7 years; mean duration of the disease 7.4 years, standard deviation 5.6 years and mean duration of levodopa therapy 5.4 years, standard deviation 3.3 years. The classical Parkinson triad: tremor, rigidity and akinesia was assessed with a three-point rating scale. Also ADL functions were rated on a three-point scale for walking, dressing, eating, washing and talking; and in addition the patients were asked about the ability to turn over in bed, which was also scored. Walking was scored as the number of steps over a distance of 6 m. The patient's subjective opinion about his or her state was marked on a 100-mm line.

In order to determine abnormalities in voluntarily induced automatic behaviour we studied posture and postural changes. Thus we examined: turning, i.e. rolling over on a flat surface; rising from a horizontal position; walking (stance and arm-swing); postural fixation (in a sitting position partly flexed for 3 min) and corrective postural reflexes, the latter with a tilting chair. Again a three-point rating scale was used. Protective postural reflexes, i.e. reflexes elicited when the body loses its balance, were discarded on account of practical reasons. (Full details of this study are published in Lakke et al. 1982.)

Results

The relationship between these parkinson symptoms was analysed with Spearman rank correlations, see Table 1. The matrix shows that all intercorrelations are generally low, some reach the 5% significance level. This indicates that the various symptoms can occur with a great deal of independence.

Next we analysed the principal components of the parkinsonian symptom intercorrelations, which suggest the following clusters (Fig. 1):

1. Self-rating, akinesia, rigidity, postural fixation and locomotor reflexes (stance and arm-swing)
2. Tremor
3. Turning both by testing and questionairre righting, walking (number of steps) and ADL function

The reaction on tilting, i.e. the corrective postural reflexes, takes a somewhat separate place which can not be explained at this time. The first cluster may be characterized by posture (maintenance of balance), the third cluster pertains to postural changes. With regard to turning and righting, clinical experience has shown (Lakke et al. 1980) that these motor abilities deteriorate despite L-dopa therapy and moreover are frustrated by physiotherapy. The motor activity in the cluster of postural changes can be characterized as voluntarily induced automatic movements, where balance plays a conditioning role.

Table 1. Spearman rank correlation between parkinsonian symptoms

	Tremor	Rigidity	Akinesia	Self-rating	Walking (stance, arm-swing)	Postural fixation	Corrective, postural, reflexes	Turning (testing)	Turning (ADL)	Righting	Walking (number of steps)	ADL
Tremor												
Rigidity	0.30											
Akinesia	0.34*	0.25										
Self-rating	0.08	0.24	0.20									
Walking (stance, arm-swing)	0.26	0.36*	0.44*	0.35*								
Postural fixation	0.00	0.25	0.27	0.37*	0.14							
Corrective, postural, reflexes	0.16	0.21	0.21	0.02	0.19	0.02						
Turning (testing)	0.13	0.38*	0.25	0.06	0.44*	0.06	0.28					
Turning (ADL)	0.20	0.36*	0.13	0.21	0.19	0.08	0.21	0.35*				
Righting	0.10	0.39*	0.20	0.30	0.28	0.15	0.23	0.31	0.48*			
Walking (number of steps)	0.13	0.05	0.24	0.21	0.39*	0.15	0.07	0.38*	0.36*	0.51*		
ADL	0.14	0.11	0.24	0.16	0.30	0.02	0.21	0.22	0.42*	0.49*	0.33*	

* Significant at $P = 0.05$ (two-tailed).

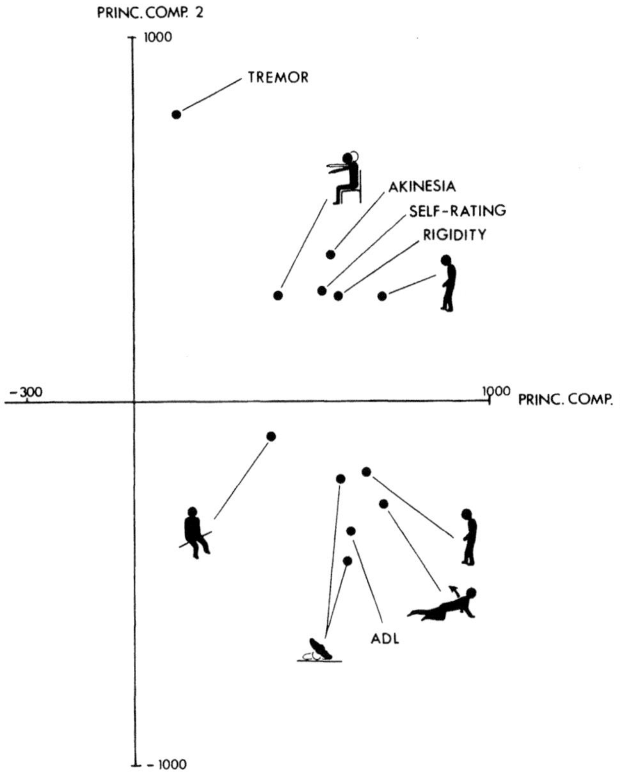

Fig. 1. The general structure of the intercorrelation of parkinsonian symptoms explored with a principal components analysis on the Pearson correlation matrix. The clustering is explained in the text

It is noteworthy that the patients themselves tend to appreciate akinesia, rigidity and the static postural reflexes as of more importance for their subjective well-being than motility, which is reflected by the fact that self-rating is found in this very cluster. In contrast, in the cluster of postural changes, which do not respond favourably on L-dopa therapy, the ADL functions are included. This finding is in accordance with the experience of Franklyn et al. (1981) that the physiotherapeutically treated patients report subjective improvement, while the "more objective" ADL functions do not parallel the patient's opinion.

Observation

In particular, we turned our attention to turning, i.e. axial rotation in a recumbent position, because in recumbency, gravitational forces, and thus gravity-dependent postural reflexes, are to a great extent inoperative. Actually in maintaining the supine position no activity is needed. Video recordings were made of rolling over on a flat sur-

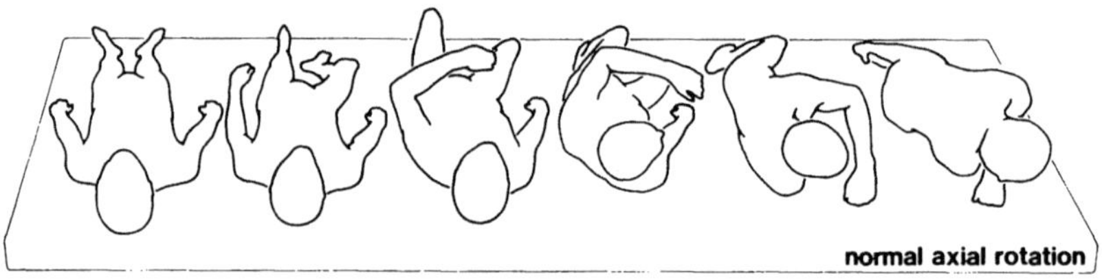

Fig. 2. Axial rotation in recumbent position of a normal person. Bending of the neck with a slight sideways rotation of the head is followed by arm-swing, bending of one knee and then rotation of the trunk. Drawings taken from sequential photographic picture frames

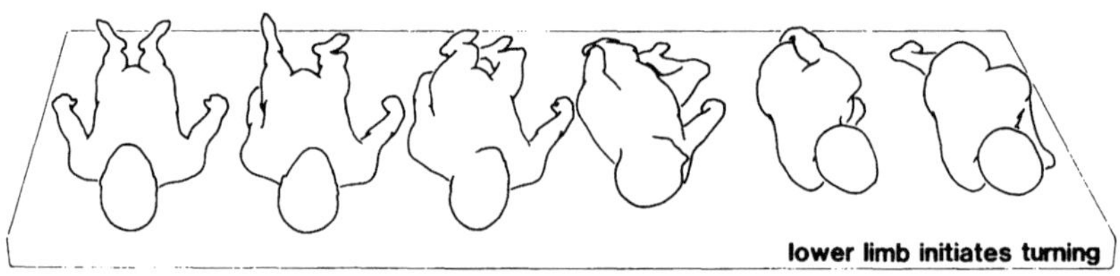

Fig. 3. The first abnormality in axial rotation appears to be an absent arm-swing. Lower limbs are bent and initiate turning of the trunk

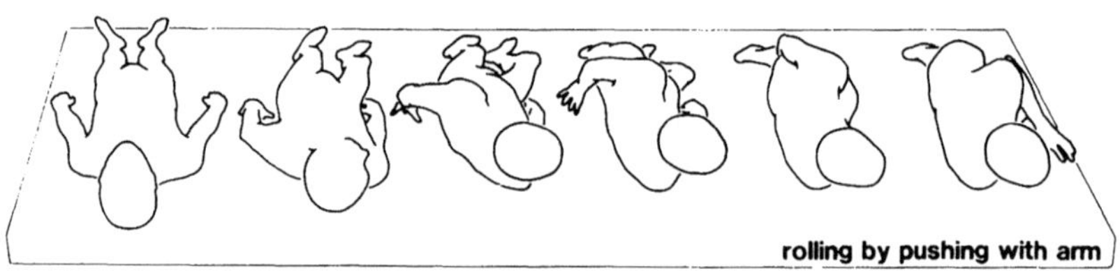

Fig. 4. Rotation is aided by pushing the body around with the hand

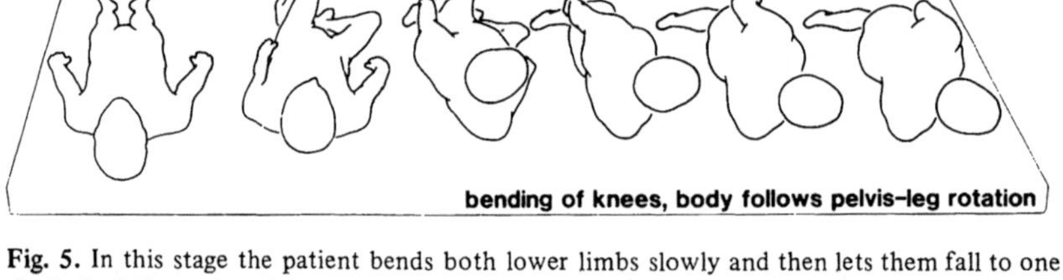

Fig. 5. In this stage the patient bends both lower limbs slowly and then lets them fall to one side; the body follows the pelvis and leg rotation

face; normal individuals were studied, a control group of elderly patients admitted to the department of ophthalmology (persons with marked disturbance of vision were excluded) and parkinsonian patients. The persons were asked to assume the supine position and then requested to roll over to the left and to the right. The healthy individuals and the control group turned according to a rather stereotyped pattern, as depicted in Fig. 2, although the elderly persons usually turned more slowly.

In the motor behaviour of parkinsonian patients we observed a variety of turning abnormalities (Figs. 3, 4, 5). Qualitative analysis of these turning patterns leads to the suggestion that axial rotation is deteriorating in stages. First, the arm-swing is gradually abolished, but rolling is in general preserved, initiated by bending of the knees. The next step seems to be that turning is purposely supported by pushing over with one arm. The next stage seems to involve flexing both legs and allowing the legs to fall sideways while the body follows. Finally, rolling over becomes impossible and only a slow forward bending of the head is performed and some restlessness of the legs can be observed. These observations suggest that the deterioration of the motor programme of axial rotation runs a course from distal to axial (from lateral to sagittal) and first loses adequate synergic action of the more peripheral musculature. Thus the ability of automatic laterality disappears, but a conscious programmatic conception of axial rotation aids in a voluntary effort with the arm. Then only flexion of limbs and axial musculature is possible and finally only some form of bending in the sagittal plane is preserved. It seems possible that a comparable pattern of deterioration occurs in walking: first the arm-swing disappears, then turning while walking becomes difficult, the stooped posture increases and festination finally incapacitates the patient.

Electromyography. One of my colleagues, Dr. E.H. Koppejan, made polymyographic recordings of axial rotation in the recumbent position, using surface electrodes. Normal persons, both young and elderly, and parkinsonian patients were studied. His findings, only qualitatively analysed, indicate that in normal individuals there is a programmed concurrent and sequential contraction and relaxation of muscles as could be expected. The recordings showed that the movement is initiated by turning of the eyes in the direction of turn, followed by bending of the neck and turning of the head. Before starting the actual axial rotation a ventromedial movement of the shoulders and an inclination of the pelvis by contraction of the abdominal musculature causes the necessary rounding of the back, to disrupt the stable supine position; then rolling over can ensue. In parkinsonian patients various responses to the request to turn were observed. Some patients were able to turn in a manner comparable to the controls, only slower. Likewise a sequential but delayed pattern of contraction and relaxation of musculature was recorded. In patients who could not turn a chaotic EMG pattern was seen, but it was found that these patients were able to execute these movements at another time almost normally with the stereotyped EMG response. This suggests that the motor programme for the execution of these postural changes is preserved, but is not automatically accessible to the patient.

Discussion

It is hazardous to suggest a more concise terminology for the postural changes we studied in the context of motor behaviour, because the physiological organisation of motility and the role of the various brain structures are only partly known. The arrangement of movements in hierarchical levels, a concept of John Hughlings Jackson (1835-1911) and still accepted (Phillips and Porter 1977), ascending from "most automatic" to "least automatic" recognized that in almost all motor activity an automatic component has to be present. In our study the term voluntarily induced motor behaviour pertains to postural changes, basic movements for displacement, acquired very early in life and considered by some as part of postural mechanisms (Purdon Martin 1967). From our data and clinical experience we conclude tentatively that voluntarily induced automatic postural displacements have the following characteristics:

1. They are to be distinguished from learned motor programme for skilled motor behaviour, because the latter do not deteriorate during L-dopa treatment.
2. In the hierarchy of postural mechanisms, these kinetic postural movements form a separate group, distinct from "static" postural reflexes.
3. Other than dopaminergic transmitters are involved.
4. Despite the disease process, the postural motor programmes can be "retrieved" under certain circumstances and thus remain functionally intact.
5. The ability to use these motor programmes deteriorates according to a certain pattern.
6. In basal ganglia disorders, specific physiotherapeutic training is unable to restore normal postural movement.

There is sufficient evidence to assume that the basal ganglia can be regarded as neural substrates for motor control and that these structures are involved in the automatic execution of learned motor plans (Marsden 1982). In parkinsonism, whereby nigrostriatal dopaminergic deficiency causes inhibitatory-exitatory imbalance in the neuronal loop organization of the basal ganglia complex, motor behaviour is largely disordered. Learned motor plans, possibly including all willed motor behaviour, comprises various categories, not yet properly defined. Indeed, L-dopa therapy improves skilled movements and part of the postural functions. However, our findings suggest that automatic postural displacements (to be considered at least partly to depend upon learned motor plans) are a category to be distinguished from skilled movements and static postural mechanisms. The neuroanatomical substrate may be outside the territory of the basal ganglia and other than dopaminergic transmitters may be involved. The disappointing results of physiotherapy may be explained by the supportive of normal motor behaviour and inaccessibility of these for postural displacement programmes by the fact that our conception of the physiological organization of motor function is still generally incomplete.

References

Franklyn S, Kohout LJ, Stern GM, Dunning M (1981) Physiotherapy in Parkinson's disease. In: Clifford Rose F, Capildeon R (eds) Research progress in Parkinson's disease. Pitman, London, pp 397-400

Lakke JPWF, de Jong PJ, Koppejan EH, van Weerden TW (1980) Observations on postural behaviour: axial rotation in recumbent position in Parkinson patients after L-Dopa treatment. In: Rinne UK, Klinger M, Stamm G (eds) Parkinson's disease. Current progress, problems and management. Elsevier/North-Holland, Amsterdam, pp 187-196

Lakke JPWF, van den Burg W, Wiegman J (1982) Abnormalities in postural reflexes and voluntarily induced automatic movements in Parkinson patients. J Clin Neurol Neurosurg 84: 227-235

Marsden CD (1982) The mysterious motor function of the basal ganglia: the Robert Wartenberg Lecture. Neurology (NY) 32: 514-593

Phillips CG, Porter R (1977) Corticospinal neurones. Their role in movement. Academic, London

Purdon Martin J (1967) The basal ganglia and posture. Pitman, London

Significance of Mental Disorders in the Long-Term Treatment of Parkinson's Disease

P.-A. Fischer

Introduction

Ample reference is made in the literature to mental disorders in parkinsonian patients (Hauptmann 1922; Jacob 1955; Mindham 1970; Ajuriaguerra 1971; Carron et al. 1972; Celesia and Wanamaker 1972; Bottenberg 1973; Jacobi et al. 1974; Danielczyk 1978; Lieberman et al. 1979; Presthus 1980; Fischer 1982). The turning point in antiparkinson treatment came with the introduction of L-dopa, which coincided with a change in the epidemiologic situation overall. After the decline in postencephalitic Parkinson's disease, attention focused on the idiopathic parkinsonian syndrome, with its manifestation at age 60 years. The disease, as it is encountered today, is one of old age and, at the same time, one that frequently implies mental disorders. Besides the degenerative nigrostriatal process with its neurotransmitter disorders, we must consider the possibility of senescent brain changes and independent age-related diseases affecting the outcome of therapy.

Subjects and Methods

We started systematic long-term studies on the therapeutic efficacy of L-dopa treatment in parkinsonian patients in 1970. To date, we have followed up approximately 350 patients for 1-12 years, all of whom have been subjected to the same examination procedure and undergone regular checkups after stabilization. Questions about the significance of mental disorders during the long-term therapy will be answered by different subgroups of this material.

Our studies were conducted from the beginning in a multivariate design, allowing for evaluation of subjective complaints, specific neurologic symptoms, identical exploration and neuropsychologic diagnostics. In order to detect possible brain dysfunctions and extranigral morphological brain changes, tests were regularly combined with EEG recordings and, after its introduction in 1974, with computer tomography (for methology see Fischer et al. 1971, 1978).

Results

Long-Term Course of the Parkinsonian Syndrome During Treatment

The reduction of subjective complaints, the amelioration of the cliniconeurologic status, and the substantial increase in psychomotor test performance indicate a marked improvement in the parkinsonian syndrome under L-dopa therapy, which is fairly consistent over a period of 2-3 years. Around the third year, however, there is a substantial evidence that efficacy starts to decline. From then on, neurologic findings deteriorate gradually, especially the target symptoms of akinesia and rigidity. In terms of mean performance level, it seems that ability deteriorates to baseline after 10 years of treatment. This generalized analysis of neurologic symptoms naturally disregards differences in the disease course, which can only be detected in individual evaluations. From a purely descriptive point of view, it seems that 90% of our case material is easily classifiable by neurologic criteria into four course types over a 5-year period. We have designated these types by the letters L, V, M and U (Fischer et al. 1980). The two extremes L and V account for about 15%-20% of the case material. Under therapy, L improves markedly in the stabilization phase and there is improvement throughout the period of observation. V gives a less positive response in the first phase of treatment; it subsequently returns more rapidly to baseline, which is eventually substantially exceeded. This analysis of events underlines the need for bringing the observed improvement in neurologic syı ptoms into line not only with general group behaviour but with each course type.

Depression in Patients with Parkinson's Disease

Upon entry and prior to initiating treatment, 25.6% of 285 patients were subdepressive or depressive by any clinical rating. In the course of treatment, the mood of these patients improved subjectively upon inquiry and objectively after checking the symptoms. By the third year, the depressive symptoms re-emerge. When investigating the correlation of "depressed mood" and other characteristics, we found it was statistically significant for akinesia and rigidity, and almost significant for sex, i.e., there was a greater incidence of depression among women.

During the first 2 years of treatment, when parkinsonian symptoms improve, we find very little correlation with neurologic symptoms, but significant correlation with sociodemographic data. The sex factor gains importance, and marital status and age come to play a role. Thus, single parkinsonian patients are more inclined to be depressive than married ones. Among singles, recently widowed patients are more seriously affected than those living alone for quite some time. As treatment progresses, however, there is an interesting change: the number of patients with depressed mood and associated akinesia, especially, rigidity, increases. After 3 years of treatment, a new correlation appears, namely between depressed mood and intelligence test ratings. Patients with more pronounced intellectual disability are more frequently found in the depressive group.

Great discrepancies exist between mood and severity of symptoms in individual cases, even when parkinsonian symptoms are positively influenced. Mood and neurologic symptoms can be directly opposed to each other (Fischer et al. 1982).

Intellectual Disorders in Parkinson's Disease

Intellectual disorders are fairly common among parkinsonian patients. The incidence of dementive changes reported in the literature varies between 14% (Hoehn and Yahr 1967) and 81% (Martin et al. 1973). This indicates that different criteria are being applied in the rating of intellectual disability. Upon closer analysis, it appears that the differences pertain primarily to the milder forms of intellectual impairment. The percentage of marked to moderate dementive changes by contrast, is usually 20%-30% (Schneider et al. 1982). The frequency of mental disorders that is generally anticipated in parkinsonian patients and the possibility of a concurrence of alterations that are specifice to Parkinson's syndrome with senescent brain changes and cerebrovascular disorders induced us to undertake cranial computed tomographic (CT) analyses.

After the introduction of cranial computer tomography in 1974, we were the first group to carry out regular examinations of patients with Parkinson's disease (Fischer et al. 1976). We recorded pathologic findings in approximately 50% of these patients. This percentage has increased somewhat since the introduction of the second generation of CT scanners. The most common extranigral findings was cerebral atrophy, varying localization and extent. The etiology of these alterations is certainly not consistent. Concomitant progressive cerebral atrophy, cerebrovascular atrophy, and residual cerebral lesions must be considered. A global comparison of patients with and without cerebral atrophy revealed that during the first 2-3 years of long-term therapy patients with pathologic CT findings do not respond as well to L-dopa therapy (Fischer et al. 1976; Schneider et al. 1978, 1979). With regard to the correlation of individual CT findings with an unfavourable course of the disease, we were able to find that evidence of cerebellar atrophy with concurrent general atrophy is unfavourable for the prognosis. The same is true of severe vascular alterations, such as hypertensive encephalopathy. Further CT examinations during the course of the disease show an increase in cerebroatrophic alterations within the first 2 years in approximately 23% of cases. This group of patients differs from the others in that the patients are older and show more dementive alterations during exploration and in psychological test results (Schneider et al. 1979). In our intelligence tests, we followed the concept introduced by Horn and Cattel (1963) of crystalline and fluid intelligence. An example of tests measuring crystalline intelligence is the Vocabulary Test forming part of the Hamburg-Wechsler test. The nonlinguistic figure reasoning test (Daniels 1971) assesses fluid intelligence. Over several years of monitoring, crystalline intelligence was found unimpaired in parkinsonian patients. Fluid intelligence, by contrast, showed a steady deterioration after temporary improvement in the initial phase of treatment. Older patients receiving chronic treatment are more frequently affected.

When determining the general level of cerebral organic ability, we found the measurement of complex visual reaction time quite useful (Fischer et al. 1980). Besides the basic speed of response to a stimulus, we measured central decision time in response to a dimensional stimulus, in addition to the time required for its more complicated motor reaction. We found a negative impact of the parkinsonian syndrome on reaction time. In studies of complex reaction time in patients with other brain diseases, reaction time was also found to be prolonged. In untreated parkinsonian patients, the possibility thus exists that a pathologic reaction time is the result of either a specific parkinsonian syndrome or an independent, extranigral brain dysfunction, or both Changes in reaction time during antiparkinson treatment would then distinguish between the two, since only parkinsonian symptoms respond to treatment. No improvement, or only slight improvement, in complex reaction time during the first phase of treatment may be an early indicator of the later outcome of treatment. This of course applies only to patients with pathologic reaction time before treatment, as we retrospectively found when classifying our material into good and poor responders. Strikingly, despite the fact that both groups differed very little in their neurologic pretreatment status, those with better reaction times tended to have the more severe parkinsonian symptoms. After a 1-year treatment, complex reaction time improved in the group of good responders by 28%, and in the negative group by only 9%. The results of reaction time determinations in parkinsonian patients are statistically significantly related to EEG findings. Patients with below average reaction times have more than coincidental decreases in alpha activity and increased theta and delta waves. Elderly parkinsonian patients with more severe intellectual impairment, with an increase in slow EEG waves, and CT evidence of brain lesions respond more poorly to treatment than do others. Symptomatic improvement during the stabilization phase and influence on complex visual reaction time may indeed serve as a measure of prospective therapeutic success. This is in full agreement with the four types of courses devised by us on the basis of overall symptoms, whose favourable or unfavourable outcome is primarily determined by age at onset and the magnitude of psycho-organic changes before therapy begins.

Drug-Induced Psychosis in Parkinson's Disease

Before the introduction of L-dopa and dopamine agonists several reports had been published on exogenous psychoses. In the past few years, it seems that the incidence of exogenous psychoses during therapy is increasing. Rates of 20%-30% are in the majority. We found exogenous psychoses in 28% of our patients. The most frequent manifestation was confusion. Since psychoses of this type are often observed under L-dopa treatment, they were called dopa psychoses, but we prefer the designations "drug-induced parkinsonian psychosis" and "pharmacotoxic psychosis". In 84 patients receiving anticholinergic monotherapy, we found an 11% incidence of psychotic episodes. When treatment was replaced by L-dopa monotherapy, all patients but one experienced psychotic episodes. We do observe exogenous psychoses in patients receiving different single-agent or combination treatments. Those with psychotic episodes are usually in the older age group when the disease strikes and late in receiving treatment. They also have more pronounced psycho-organic changes. This by itself confirms the poor prognosis previously refered to from a pharmaceutical point of view. For practical reasons, we should like to mention, further, that 50% of parkinsonian patients with exogenous psychotic episodes have had concomitant diseases or other debilitating factors (e.g., operations) at the time of onset. Congestive heart failure requiring treatment is of particular importance in this respect. But usually factors other than a central drug action are also involved. Thus, patients with this poor prognosis frequently suffer from a second disease, are fairly old, and undergo chronic treatment. Exogenous psychotic episodes are, thus, primarily a diagnostic problem.

Conclusions

The examination results presented here underline the importance of mental and psychic disorders in the treatment of parkinsonian patients. The results of our long-term studies have shown that some concomitant psychic symptoms, such as dementia, are much more likely to affect prognosis than is the severity of the neurologic symptoms upon initiation of treatment. Amenability, course, and prognosis of the parkinsonian syndrome will be determined to a considerable extent by cerebral dysfunction and its morphological implications, above and beyond the specific alterations of the nigrostriatal system.

The search for improvements in antiparkinsonian therapy should of necessity imply a detailed analysis of accompanying extranigral cerebral disorders, for which psychopathologic findings are important correlates.

References

Ajuriaguerra J de (1971) Etude psychopathologique des parkinsoniens. In: Ajuriaguerra J de, Gauthier G (eds) Monoamines noyaux gris centraux et syndrome de parkinson. Georg, Geneva

Bottenberg EH (1965) Zur Psychodiagnostik des Parkinson-Syndroms. Methoden und Ergebnisse. Fortschr Neurol Psychiatr 41: 237-282

Carron DC, Klawans HL, Narim F (1972) Intellectual functioning of persons with idiopathic parkinsonism. J Nerv Ment Dis 154: 445-452

Catell RB (1963) Theory of fluid and crystallized intelligence: a critical experiment. J Educ Psychol 54: 1-22

Celesia GG, Wanamaker WM (1972) Psychiatric disturbances in Parkinson's disease. Dis Nerv Syst 33: 577-583

Danielczyk W (1978) Akute psychische Störungen während der L-Dopa-Therapie von Parkinson-Kranken. In: Fischer PA (Hrsg); Langzeitbehandlung des Parkinson-Syndroms. Schattauer, Stuttgart, pp 211-218

Daniels JC (1971) Figure reasoning test: Instruction manual. London

Fischer PA (ed) (1982) Psychopathologie des Parkinson-Syndroms. Roche, Basel

Fischer PA, Jacobi P, Schneider E, Becker H (1976) Correlation between clinical and CT findings in Parkinson's syndrome. In: Lanksch W, Kazner E (eds) Cranial computerized tomography. Springer, Berlin Heidelberg New York, pp 244-248

Fischer PA, Schneider E, Jacobi P (1978) Die Langzeitbehandlung des Parkinson-Syndroms mit L-Dopa. Befunde und Probleme. In: Fischer PA (ed) Langzeitbehandlung des Parkinson-Syndroms. Schattauer, Stuttgart, pp 87-103

Fischer PA, Schneider E, Jacobi P (1980) Verlaufsformen und Verlaufsfaktoren beim Parkinson-Syndrom. In: Fischer PA (ed) Parkinson-Syndrom: Kombinations- und Begleit-Therapien. Schattauer, Stuttgart, pp 97-115

Fischer PA, Schneider E, Jacobi P (1982) Depressive Verstimmungen bei Parkinson-Kranken im Langzeitverlauf. In: Fischer PA (ed) Psychopathologie des Parkinson-Syndroms. Roche, Basel, pp 139-152

Hauptmann A (1922) Der "Mangel an Antrieb" von innen gesehen (das psychische Korrelat der Akinese). Arch Psychiatr Nervenkr 66: 615-687

Hoehn MM, Yahr MD (1967) Parkinsonism: onset, progression, and mortality. Neurology (NY) 17: 427-442

Jacob H (1955) Wahrnehmungsstörung und Krankheitserleben. Psychopathologie des Parkinsonismus und verstehende Psychologie Bewegungs- und Wahrnehmungsgestörter. Springer, Berlin Göttingen Heidelberg (Monogr Gesamtgeb Neurol Psychiatr Heft 78)

Jacobi P, Fischer PA, Schneider E (1974) Das Verhaltn der allgemeinen Intelligenz von Parkinson-Patienten während der L-Dopa-Therapie. Arch Psychiatr Nervenkr 218: 125-140

Liebermann A, Dziatolowski M, Kupfersmith M, Serby M, Goodgold A, Korein J, Goldstein M (1979) Dementia in Parkinson's disease. Ann Neurol 6: 355-359

Mindham RHS (1970) Psychiatric symptoms in parkinsonism. J Neurol Neurosurg Psychiatry 33: 188-191

Presthus J (1980) Psychiatric side effects occurring in parkinsonism during long-term treatment with levodopa alone and in combination with other drugs. In: Rinne UK, Klinger M, Stamm G (eds) Parkinson's disease – current progress, problems and management. Elsevier/North-Holland, Amsterdam, pp 225-270

Schneider E, Fischer PA, Jacobi P, Becker H (1978) Zur Relevanz extranigraler Hirnläsionen bei Parkinsonkranken. In: Fischer PA (ed) Langzeitbehandlung des Parkinson-Syndroms. Schattauer, Stuttgart, pp 115-129

Schneider E, Becker H, Fischer PA, Grau H, Jacobi P, Brinkmann R (1979) The course of brain atrophy in Parkinson's disease. Arch Psychiatr Nervenkr 227: 89-95

Schneider E, Fischer PA, Jacobi P (1982) Demenz beim Parkinson-Syndrom. In: Fischer PA (ed) Psychopathologie des Parkinson-Syndroms. Roche, Basel, pp 93-114

Treatment of Parkinson's Disease in Early and Late Phases

M. Yahr

Long-Term Effectiveness of L-Dopa

In 1975 at the 5th International Symposium on Parkinson's disease held in Vienna under the Presidency of Professor Birkmayer, we reported on the long-term results of L-dopa administration. In some 600 patients who had entered treatment at various phases of the disease and had been receiving dopa for as long as 7 years, the results depicted in Fig. 1 were obtained.

An optimal therapeutic response occurred during the first 3 years of drug usage with some 80% of patients showing an average improvement of 70%. Following this there was gradual decline in therapeutic efficacy decreasing to only 40% improvement. The first obvious conclusion from these data was that, with progression of the disease, the patients became more unresponsive to therapy. However, more detailed study in which cases were separated in regard to the degree of severity on entering treatment, suggested an alternative reason (Fig. 2). All patients regardless of severity of disease, excluding those in the very severe end stage, responded initially and then in similar fashion showed a declining response. In view of this, it became apparent that one had to consider the drug itself as playing a role in a loss of efficacy. There is at present substantial evidence that this is so. The report of Lee et al. (1978) on postmortem parkinson brains in which those untreated with dopa drugs showed evidence of dopaminergic receptor supersensitivity while those treated with such agents had lost such a response is compelling evidence of a pharmacologic dynamic action with resultant morphological change due to administration of dopaminergic agents, such a change can reduce effectiveness of L-dopa. Further, from a clinical standpoint there is evidence of increasing incidence of side effects which occur over time and limit the beneficial effects of dopa. Induced involuntary movements became an increasing problem and no effective means for their control without giving up some of the benefits has been devised. We have used an extensive number of agents without success in this regard (Table 1). Even more distressing are the fluctuating responses, so called on-off reactions, for which no fully satisfactory explanation or treatment exists. The number of approaches now being uses in an attempt to control these (Table 2) is indicative of the fact that we are still at sea as to its true mechanism of action. Some degree of control of these distressing reactions can be obtained by one or another, but none are applicable to all patients.

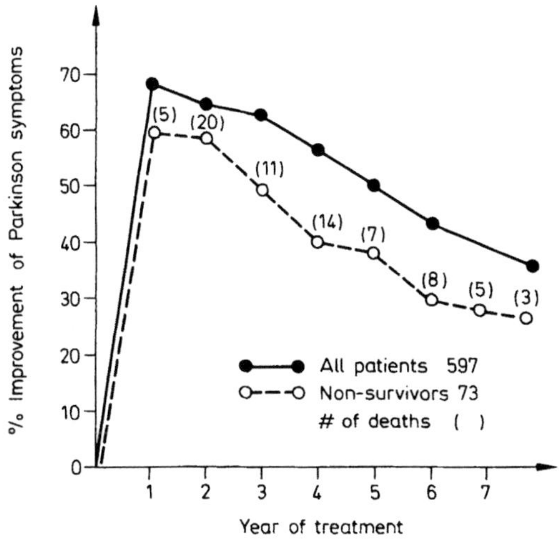

Fig. 1. Degree of improvement of parkinsonian symptoms during 7 years of chronic L-dopa treatment. Yahr (1975)

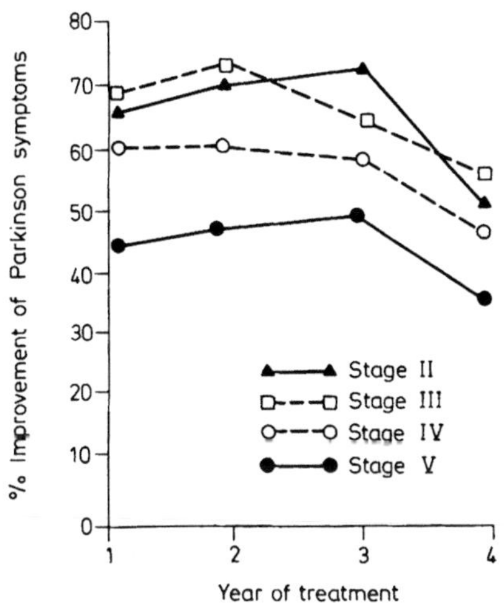

Fig. 2. Comparison of response to chronic L-dopa therapy in groups of parkinsonian patients with different degrees of severity

Pharmacologic Mechanisms in On-Off Responses

Some interesting features have been accumulated regarding on-off responses and I should like to touch on a few. The first concerns what might be termed an altered metabolism of exogenously administered L-dopa leading to unstable plasma levels. In some patients the ability to maintain consistently elevated plasma L-dopa levels after long-term usage is altered. As indicated in Fig. 3, a patient suffering from the on-off response could only maintain an effective plasma level of L-dopa for a short interval after an ingested dose. More recently, we have been able to show that the concept that dopaminergic receptors become depolarized and temporarily inactive during an off phase is untenable, as in-

Table 1. Drugs administered for control of abnormal involuntary movements (AIM)

	AIM	*Parkinsonism*		*AIM*	*Parkinsonism*
Anti-ACh	0	↓	Dilantin	0	0
Stelazine	↓	↑	Barbiturates	0	0
Haloperidol	↓	↑	Reserpine	↓	↑
Imipramine	0	↓	Clonazepam	0	0
Disulfiram	↑	↑	Di-*n*-proplylacetate	0	0
Pyridoxine	↓	↑	Deanol	0	↓
L-Tryptophan	0	0	Lithium	0	0
L-5-HTP	0	0	Choline	±	↑
L-Methionine	0	0	Progabide	±	↓
Parachlorophenylalanine	0	0			

Table 2. Therapeutic measures useful in treatment of on-off reaction

1. Maintain established L-dopa dosage – alter dosing schedule using multiple small doses
2. Increase total daily dosage of L-dopa with or without change in dosing schedule
3. Discontinue L-dopa (7-14 days) – reinstitute starting with lower dosage and gradually increase (drug holiday)
4. Use of dopamine "expanders": Amantadine HC1; tricyclic compounds; MAO-B inhibitor (deprenyl)
5. Use of dopaminergic receptor agonists: Peribedil; bromocriptine; pergolide

dicated in Fig. 4. Injection of apomorphine during an off phase reverses the parkinsonian state and indicates the availability of dopaminergic receptors for stimulation. This response is dose dependent, usually 1-2 mg of apomorphine given subcutaneously (Fig. 5). Similarly, changes appear to occur in the cholinergic system as well. The administration of physostigmine to an untreated parkinsonian patient can markedly exacerbate symptoms. When such a patient is treated with dopa with good response, injection of physostigmine does not alter the parkinsonian state (Fig. 6). However, when the on-off phenomenon ensues, sensitivity to physostigmine returns (Fig. 7). Hence, there are continuous dynamic changes in cholinergic mechanisms which I suspect may often be triggered by the type of biochemical mileu which we create by the administration of therapeutic agents.

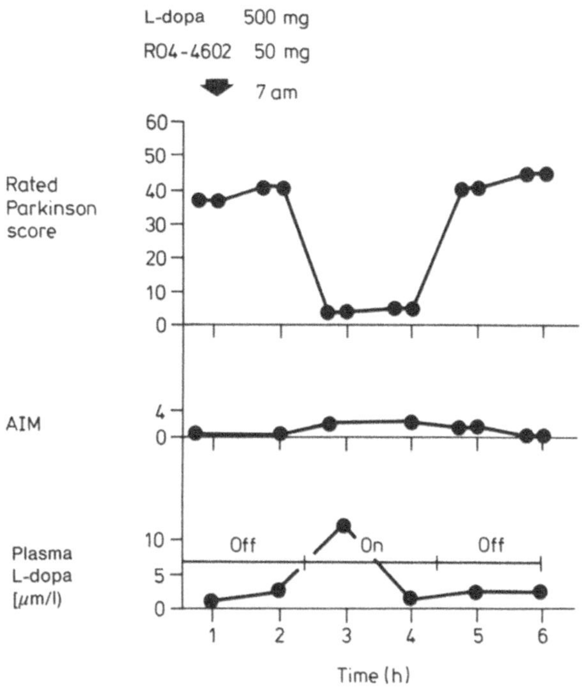

Fig. 3. Time course of Parkinson score, abnormal involuntary movements (AIM) and plasma L-dopa level after ingestion of a single dose of L-dopa plus decarboxylase inhibitor

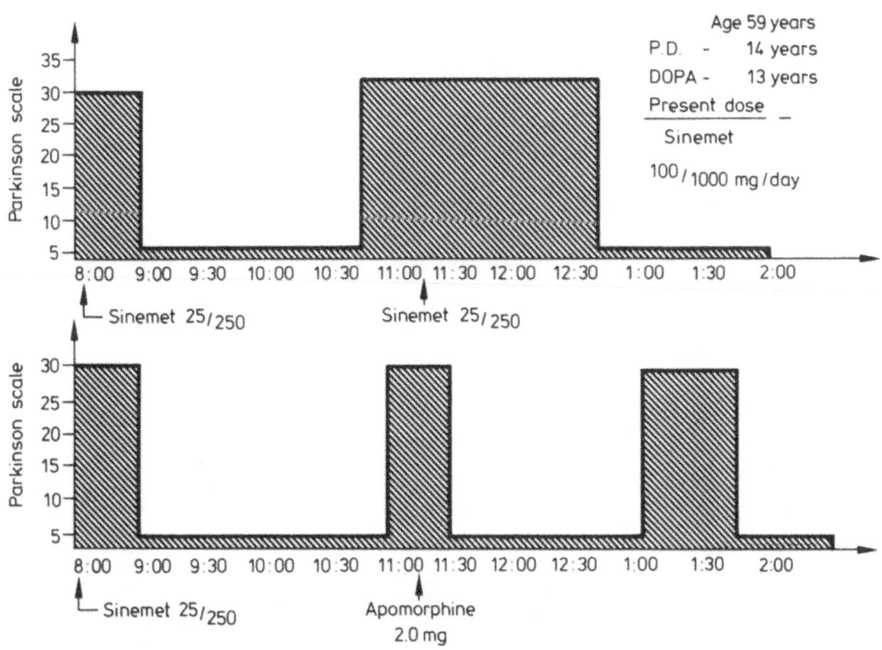

Fig. 4. Antiparkinsonian efficacy of apomorphine during off phase in a patient under chronic L-dopa treatment, indicating the presence of responsible dopaminergic receptors during off phase

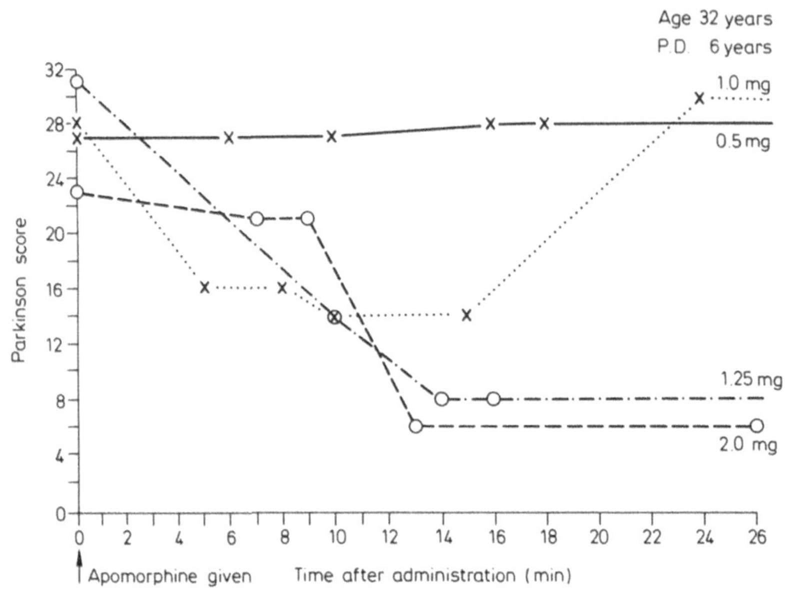

Fig. 5. Response of different subcutaneous doses of apomorphine given during off phase in a patient under chronic L-dopa treatment

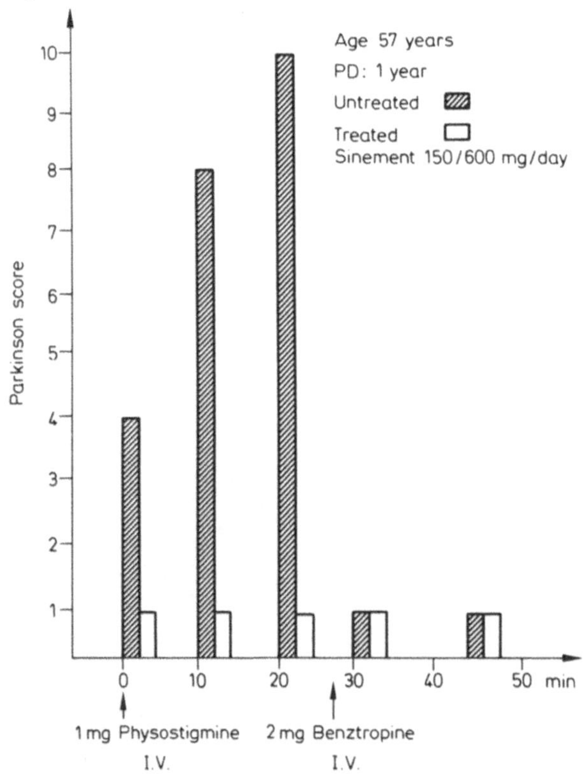

Fig. 6. Effect of intravenous physostigmine on Parkinson score of a patient before *(hatched bars)* and during L-dopa treatment *(open bars)*

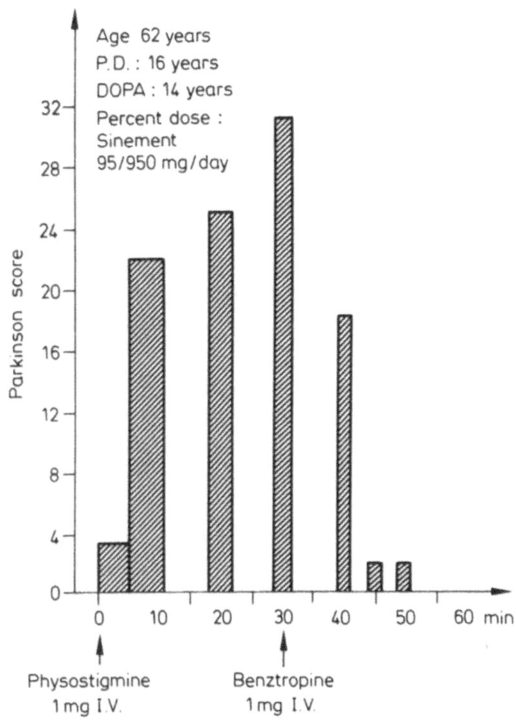

Fig. 7. Effect of intravenous physostigmine on Parkinson score of a patient with on-off reactions during chronic L-dopa treatment

Table 3. Therapeutic phases of Parkinson's disease "compensated" stages 1 and 2[a]

1.	Reassurance – no pharmacologic agents
2.	Agents with CNS antimuscarinic action: Triexyphenidyl, Benztropine, Diphenylhydramine
3.	Dopamine releasing or reuptake inhibiting agents: Amantadine HC1, Amphetamine (methamphetamine), Tricyclics (amitriptyline, imipramine)
4.	MAO inhibitor (deprenyl)

[a] Hoehn-Yahr scale.

Phases of Parkinson's Disease and Their Relation to Drug Therapy

In view of the limitations of our present pharmacological agents, and the fact that the treatment of Parkinson's disease is fundamentally symptomatic and palliative, we have suggested a practical approach to its treatment. In light of our present knowledge I think we can separate parkinsonism into two phases, the compensated and decompensated. The compensated phase is characterized by minimal symptomatology, that is, no functional impairment corresponding to stages 1 and 2 on the rating scale we suggested some years ago (Table 3). Some evidence exists that during this phase surviving dopaminergic neurons in the nigra are capable of an increased rate of neurotransmitter synthesis which in part makes up for that which has been lost as a result of the disease process. It is with the progression of the disease, when functional impairment occurs, that levodopa given with a peripheral decarboxylase inhibitor is best utilized (Table 4). I have divided this decompensated phase into two, early and late. The early phase is the interval between initiation of treatment with levodopa and the time at which the optimal therapeutic response occurs. The late phase ensues after 3-5 years of treatment when the therapeutic response diminishes and more of the complications of the use of levodopa occur.

Adjuvant Dopaminergic Agents

Let me discuss some of our experience in this late decompensated phase with agents ancilliary to levodopa and decarboxylase inhibitor. Deprenyl, a monoamine oxidase-B inhibitor has proven particularly useful in the end dose type of on-off response (Table 5). Its side effects are well tolerated though significant increase in involuntary movements occur. In regard to the ergoline derivatives, we have had experience with two of the three now available (Fig. 8). In our opinion none are as potent as levodopa itself nor are they best used alone, but are more helpful when given with dopa. In a study comparing bromocriptine and Sinemet, twice the improvement occured at the optimal dosage with Sinemet than with bromocriptine (Table 6). Bromocriptine has been particularly valuable as an adjunctive agent with Sinemet as shown in Table 7. As to pergolide (Fig. 9), which is a more potent ergoline derivative, with a particularly long half-life of some 40 h, it again is no match for the therapeutic effectiveness of Sinemet. Pergolide has properties useful in the end of dose wearing off response as shown in Table 8.

In regard to the ergoline derivatives in general, there is little doubt that they are effective antiparkinson agents, but less so than L-dopa with a peripheral decarboxylase inhibitor. Their toxicity is somewhat greater in that, in addition to those which L-dopa produces in the nervous system, others of a peripheral nature occur. Their usefulness tends to diminish in time in a similar fashion to that of L-dopa.

Conclusions

Numerous agents are presently available which are capable of controlling the symptoms of parkinsonism. None are ideal, all have their limitations. The most potent is L-dopa administered with a peripheral decarboxylase inhibitor. However, because its effectiveness diminishes after long-term use, it should be reserved for individuals with established symptoms who are functionally impaired. In those with minimal symptoms, agents which facilitate dopaminergic mechanisms normally operative in the nervous

Table 4. Therapeutic phases of Parkinson's disease "decompensated" stages 3, 4 and 5[a]

Early	1. Levodopa + PDI
	2. Levodopa + PDI + anti-ACh Dopamine release or reuptake inhibitor
Late	3. Levodopa + PDI + MAO-B inhibitor (deprenyl)
	4. Dopaminergic receptor agonists: Bromocriptine Pergolide Lisuride

[a] Hoehn-Yahr scale.

Table 5. Results of deprenyl therapy: on-off phenomena

	Number	Results[a]	
		Improved	Unimproved
End-of-dose	47	39	8
Random	19	9	10
Total	66	48 (72%)	18 (28%)

[a] Improvement defined as increase in duration of on phase by more than 50% and equivalent or improved state of parkinsonism compared with pretreatment.
Daily dosage 10 mg.
Duration of treatment 5 years.
Major side effects: AIM and mental changes.

Table 6. Comparison of bromocriptine and L-dopa + carbidopa in ten parkinsonian patients

Treatment	Parkinson score	Improvement %
None	35.2± 9.8[a]	
Bromocriptine alone	24.9±10.6	29.3
L-Dopa + carbidopa	15.4± 9.0	56.3

[a] Standard deviation.

Table 7. Benefits in 51 parkinsonian patients treated with Sinemet plus adjuvant bromocriptine

Significant clinical improvement noted in	31
On-off effect reduced	28/36
Dystonic cramps	10/10
Thermal paresthesia	1/3
Other pains	0/3
Poor response to L-dopa	0/15
Gait impairment	0/5

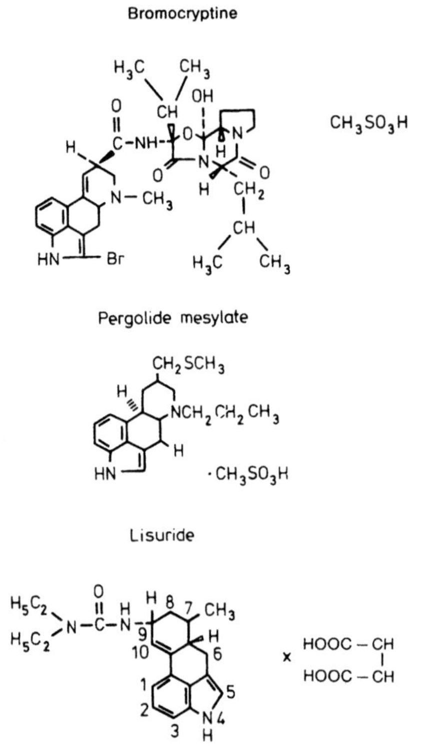

Fig. 8. Structure of three main ergoline derivatives in current clinical use

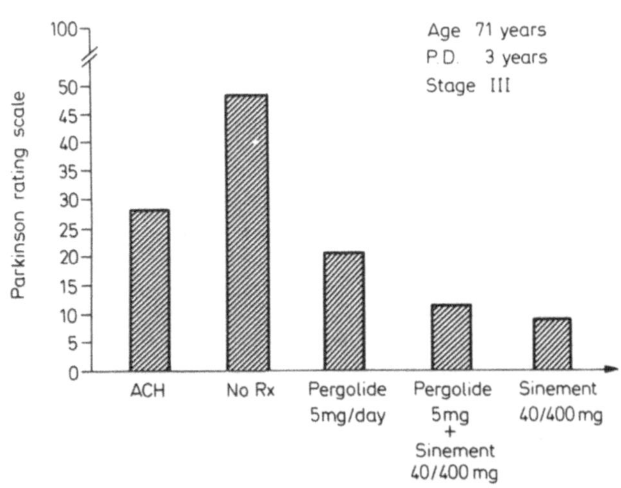

Fig. 9. Comparison of therapeutic response to anticholinergics, pergolide, pergolide plus Sinemet, and Sinemet alone in a parkinsonian patient

Table 8. Results of pergolide therapy: on-off phenomena

	Number	Results[a]	
		Improved	Unimproved
End-of-dose	18	12	6
Random	6	1	5
Total	24	13	11
		(54%)	(46%)

[a] Improvement defined as increase in duration of on phase by more than 50% and equivalent or improved state of parkinsonism compared with pretreatment.
Pergolide average daily dose 3.8 mg (2-5 mg).
Sinemet daily dose reduction 15%.
6-12 Month evaluation.
Decreased therapeutic effectiveness.
Increased average daily dose 4.4 mg/day.

system should be used. For those in the late phases of parkinsonism where L-dopa plus a peripheral decarboxylase inhibitor begins to lose its effect, combining it with a monoamine oxidase-B inhibitor or an ergoline derivative can be effective in maintaining a good functional state in many parkinsonian patients.

References

Lee T, Seeman TP, Rajput A, Farley I, Hornykiewicz O (1978) Receptor basis for dopaminergic supersensitivity in Parkinson's disease. Nature 273: 59-61

Yahr MD (1975) Evaluation of long-term therapy in Parkinson's disease: Mortality and therapeutic efficiency. In: Advances in Parkinsonism. Roche, Basel, pp 435-441

Relevance of Dopaminergic Agonists and Antagonists to Parkinsonian Tremor

S. Ruggieri, A. Denaro, M. Baldassarre, F. Stocchi and A. Agnoli

Tremor is the most complex symptom of Parkinson's disease, either for physiopathology or for neurochemistry. It seems to be the result of a derangement of cortical-subcortical loops which in Parkinson's disease also involve the nigrostriatal system (Poirier et al. 1976; Teravainen et al. 1979; Narabayashi et al. 1979). Furthermore, in tremor there are very complex modifications of neurotransmitters which cause the onset of this symptom through a derangement of the relationship of the neurotransmitters themselves (Barbeau 1962; McGeer et al. 1961; Siegfrid 1976). The nigrostriatal dopaminergic system is certainly involved in the pathogenesis of tremor, even if a clear relationship between tremor and the damage of the dopaminergic system does not exist (Rinne et al. 1982).

Treatment with drugs which restore the activity of dopaminergic neurons constantly improves tremor, the amelioration being dependent on the dosage and type of drugs used (Fig. 1). In our previous studies, this improvement ranged between 25% and 69%. In fact, for piribedil, doses of 120 and 400 mg were able to induce a 30% and 60% improvement, respectively. The highest dosage was possible by blocking peripheral effects with domperidone which did not modify tremor when administered alone (Fazio et al. 1972; Agnoli et al. 1978, 1981, 1983). This observation indicates that the different activity of drugs might be due, for the highest dosages, to a loss of receptor specificity or to the different mode of action of the drugs as releasers, receptor agonists or precursors. It might also be due to an action on other neurotransmitter systems through enzymatic competition (Pletscher 1976) on metabolism or through reciprocal interactions (Berde 1980). Furthermore, a greater receptor selectivity can be proposed: in fact ergoline derivatives, D2 agonists (Kebabian and Calne 1979; Spano et al. 1978) were the most effective on tremor (Kartzinel et al. 1976a, b; Lieberman et al. 1980, 1982, 1983; Parkes et al. 1981). Considering the therapeutic efficacy of long-term treatment with dopaminergic drugs, a study was performed to point out the relevance of D1 and D2 receptors on parkinsonian tremor.

Materials and Methods

Seven idiopathic parkinsonian patients with complete symptoms, mean age 62.6 years, six men and one woman, suffering for the disease for 6.4 years and treated for 4.1 years, in to the second and third stages of the Hoehn and Yahr classification were studied. Patients were selected because of their constant tremor without evident daily fluctuation.

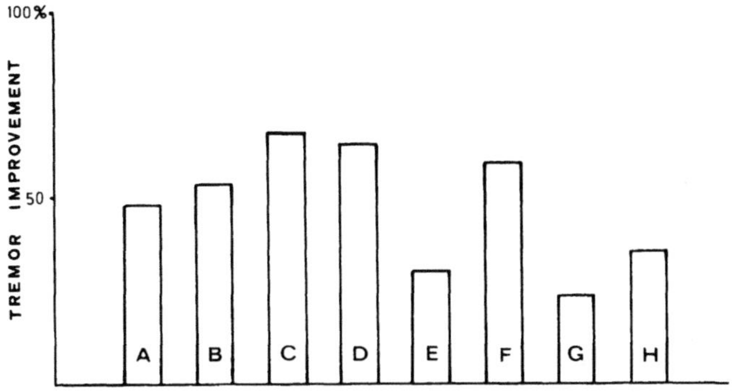

Fig. 1. Amelioration of tremor induced by treatment with different drugs: *A* L-dopa (83.5 g); *B* L-dopa+DI (750+75 mg); *C* bromocriptine (35 mg); *D* Lisuride (3 mg); *E* piribedil (120 mg); *F* piribedil + domperidone (400+45 mg); *G* amantadine (200 mg); *H* trihexyphenidyl (6 mg)

No modifications of chronic L-dopa treatment was allowed during the study. The last administration was 12 h before the acute study. Tremor was evaluated using a piezoelectric accelerometer at succeeding periods from 15 to 120 min after the injection of the drugs used. Each recording lasted 5 min and modifications of tremor were evaluated by amplitude changes. Statistical analysis was performed using Student's t-test. At time 0 the following drugs were administered: saline solution; lisuride (0.0007 mg/kg i.v.) as D2 agonist (Kebabian and Calne 1979); tiapride (2.85 mg/kg i.v.) as D2 blocker (Price et al. 1978); pentoxifylline (2.85 mg/kg i.v.) as an inhibitor of phosphodiesterase and therefore acting as D1 receptor agonist by increasing cAMP content (Goodwin-Austen et al. 1980); flupenthixol (0.014 mg/kg i.v.) as D1 blocker (Hyttel 1978); orphenadrine (1.14 mg/kg i.v.) as an anticholinergic drug whose action on tremor is well known. All drugs were brought up to 5 ml and administered at 9 a.m. at a rate of 1 ml/min.

Results

An immediate and great worsening of tremor was observed immediately after the injection of pentoxifylline which reached significance at 5-15 min. At 30 min, even if a 100% worsening is still present significance is lost (Fig. 2). On the other hand flupenthixol caused a significant improvement of tremor 60 min after injection. Even if an improvement of tremor was observed after the administration of tiapride, a D2 blocker, it remained within the values of placebo, which in these experiments range from 0% to 45% of improvement or worsening (Fig. 3). The D2 agonist lisuride showed a very important improvement of tremor at 15, 30 and 60 min and reached higher significance at 90 min. Orphenadrine caused a dramatic amelioration of tremor that was constantly highly significant during the period of observation (Fig. 4).

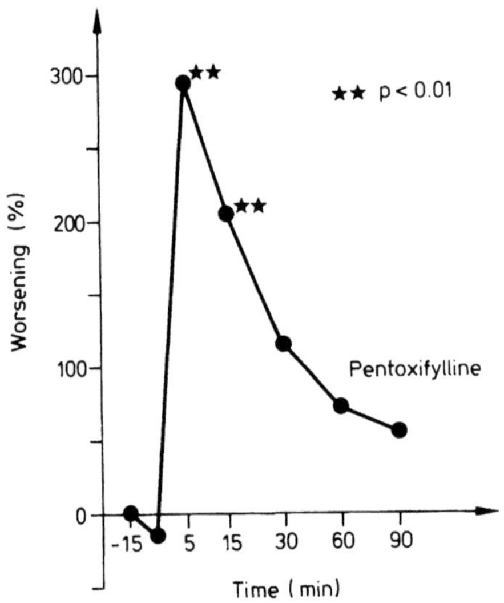

Fig. 2. Effect of intravenous pentoxifylline on parkinsonian tremor

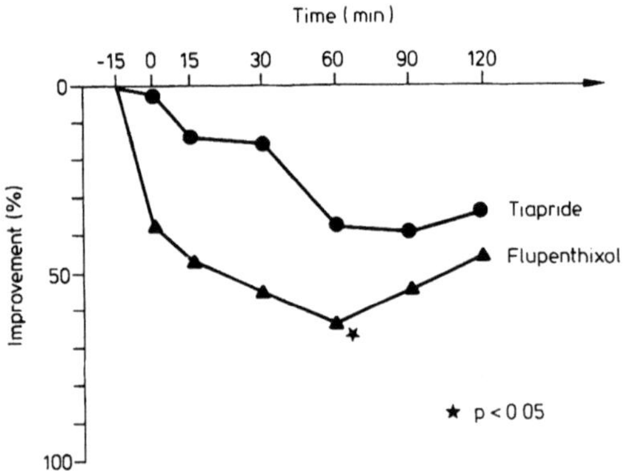

Fig. 3. Effects of parenteral tiapride and flupenthixol on parkinsonian tremor

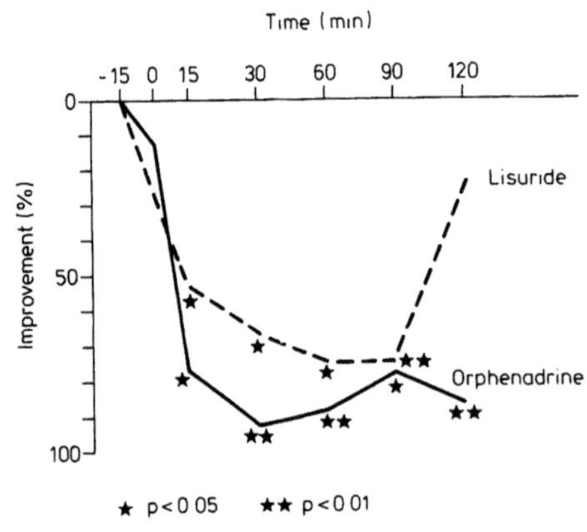

Fig. 4. Effects of intravenous lisuride and orphenadrine on parkinsonian tremor

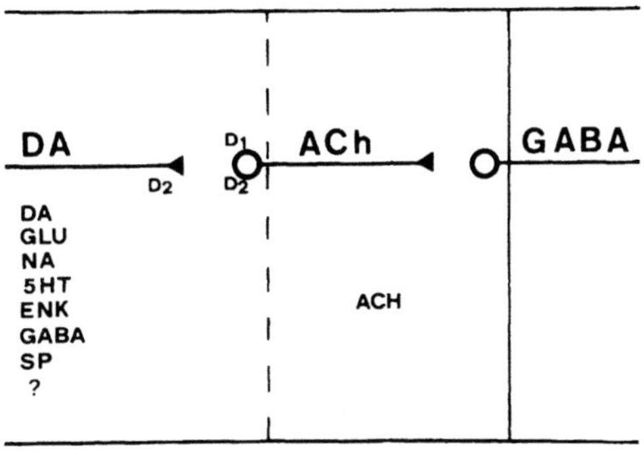

Fig. 5. Neurotransmitters which may interfere with the dopamine-acetylcholine balance (DA/ACH) at the nigrostriatal level. Anticholinergics overcome this interaction by acting directly on striofugal fibres (GABA)

Discussion

Our results show an improvement of tremor after lisuride administration while a dramatic worsening was observed after pentoxifylline. In this regard it must be underlined that phosphodiesterase inhibition does not occur only in the dopamine-dependent cAMP systems, but also in others such as the noradrenergic system, which might be involved and therefore affect tremor severity. On the other hand D1 and D2 receptor blockers were both able to improve tremor slightly.

These results indicate that the blockade of one subtype of dopaminergic receptor does not lead to a relative hyperactivity of the other, casting doubt on the hypothesis of a functional D1-D2 balance. The efficacy of dopaminergic receptor blockers might be explained by considering the sedative effects of these drugs. Moreover, dopaminergic blocking agents may be only partially selective and also exert their action on other receptors involved in tremor such as histamine and noradrenaline receptors.

The lesser improvement obtained with lisuride compared with orphenadrine might be due to the relationship existing between dopamine and other neurotransmitters which could be deranged by D2 overstimulation (Fig. 5). The system might reset with long and short feedback loops on dopaminergic neurons. On the contrary, anticholinergics overcome these interactions by blocking the receptors on efferent fibres, although an action of anticholinergics on other systems relevant to tremor cannot be excluded. The action of lisuride might be explained in the view of recent data indicating that only the stimulation of D2 receptors is able to induce a decrease of acetylcholine release in the striatum (Calne 1982).

References

Agnoli A, Ruggieri S, Casacchia M (1978) Restatement and prospectives of ergot alkaloids in clinical neurology and psychiatry. Pharmacology (Suppl 1) 16: 174-188

Agnoli A, Baldassarre M, del Roscio S, Palesse N, Ruggieri S (1981) Piribedil and Parkinson's disease: Protection of peripheral side effects by Domperidon. In: Corsini G, Gessa GL (eds) Apomorphin and other dopaminomimetics. Raven, New York, pp 117-122

Agnoli A, Ruggieri S, Baldassarre M, Stocchi F, Denaro A, Falaschi P (1983) Basic mechanisms and endocrine and neurological effects. In: Calne DB, Donald B, Horowski R, McDonald RJ, Wuttke W (eds) Lisuride and other dopamine agonists. Raven, New York, pp 407-417

Barbeau A (1962) The pathogenesis of Parkison's disease: a new hypothesis. Can Med Ass J 87: 802-807

Berde B (1980) Ergot compounds: a synopsis. In: Goldstein M, Calne DB, Lieberman A, Thorner MO (eds) Advances in biochemical psychopharmacology, vol 23. Raven, New York, pp 3-24

Calne DB (1982) Dopamine agonists and dopamine receptors: their relevance to the treatment of Parkinson's disease. International Workshop on Parkinson's disease, Igls, Austria

Fazio C, Agnoli A, Casacchia M, Reitano M, Ruggieri S, Barba C (1972) Treatment of Parkinson's disease with L-Dopa and an association L-Dopa plus a Dopa decarboxylase inhibitor. Z Neurol 202: 347-355

Goodwin-Austen RB, Twomey JA, Hanks G, Higgins J (1980) Oxpentiphylline in Parkinson's disease. J Neurol Neurosurg Psychiatry 43: 360-364

Hyttel J (1978) Effects of neuroletpics on 3H-haloperidol and 3H-cis-z-flupentixol binding and on adenyl cyclase activity in vitro. Life Sci 23: 551-556

Kartzinel R, Shoulson I, Calne DB (1976a) Studies with bromocriptine. II Double blind comparison with L-Dopa in idiopathic parkinsonism. Neurology (Minneap) 26: 511-513

Kartzinel R, Shoulson I, Calne DB (1976b) Studies with bromocriptine. III Concomitant administration of caffeine to patients with idiopathic parkinsonism. Neurology (Minneap) 26: 741-743

Kebabian JW, Calne DB (1979) Multiple receptors for dopamine. Nature 277: 93-96

Lieberman A, Kupersmith M, Neophytides A, Casson I, Durso R, Hoo Foo S, Khayali M, Bear G, Goldstein M (1980) Long-term efficacy of bromocriptine in Parkinson's disease. Neurology (NY) 30: 518-523

Lieberman A, Goldstein M, Gopinathan G, Leibowitz M, Neophytides A, Walker R, Hiesiger E, Nelson J (1982) Further studies with pergolide in Parkinson's disease. Neurology (NY) 32: 1181-1184

Lieberman A, Gopinathan G, Neophytides A, Leibowitz M, Walker R, Hiesiger E (1983) Bromocriptine and lisuride in Parkinson's disease. Ann Neurol 13: 44-47

McGeer PL, Boulding JE, Gibson WC, Foulkes RG (1961) Drug induced extrapiramidal reactions. JAMA 177: 665-670

Narabayashi M, Chida T, Kondo T (1979) Analysis of two factors influencing involuntary movements: psychological stress effect and motor effect. In: Poirier LJ, Sourkes TL, Bedard PJ (eds) Advances in neurology, vol 24. Raven, New York, pp 361-372

Parkes JD, Schachter N, Marsden CD, Smith B, Wilson A (1981) Lisuride in parkinsonism. Ann Neurol 9: 48-52

Pletscher A (1976) Biochemical and pharmacological aspects of Parkinson's syndrome: a short review: In: Birkmayer W, Hornykiewicz O (eds) Advances in parkinsonism. Roche, Basel, pp 21-36

Poirier LJ, Filion M, Larochelle L, Péchadre JC (1976) Physiopathology of tremor and rigidity. In: Birkmayer W, Hornykiewicz O (eds) Advances in parkinsonism. Roche, Basel, pp 217-235

Price P, Parkes JD, Marsden CD (1978) Tiapride in Parkinson's disease. Lancet 1106

Rinne UK, Rinne JK, Rinne JO, Laakso K, Tenovuo O (1982) Neurotransmitters and neuropeptides in the parkinsonian brain. In: Integrative Neurohumoral Mechanisms. Physiological and clinical aspects, Budapest, 24-27 March

Siegfrid J (1976) The treatment of parkinsonian tremor. In: Birkmayer W, Hornykiewicz O (eds) Advances in parkinsonism. Roche, Basel, pp 319-327

Spano PF, Govoni S, Trabucchi M (1978) Studies on the pharmacological properties of dopamine receptors in various areas of the central nervous system. In: Roberts PJ, Woodruff GN, Iversen LL (eds) Advances in biochemical psychopharmacology. Raven, New York, pp 105-115

Teravainen H, Evarts E, Calne D (1979) Effects of kinesthetic inputs on parkinsonism tremor. In: Poirier LJ, Sourkes TL, Bedard PJ (eds) Advances in neurology, vol 24. Raven, New York, pp 161-174

Therapeutic Possibilities in On-Off Phenomena in Parkinson's Disease

E. Csanda and M. Tárczy

Introduction

The therapeutic correction of on-off phenomena seems to be the major difficulty in the long-term treatment of Parkinson's disease. In order to get better information about the diurnal fluctuation of patients with on-off phenomena an hourly system for observation has been introduced. In 28 patients a number of adjuvant drugs were tested for correcting their on-off symptoms. L-Deprenyl proved to be effective in patients with the wearing-off effect. The administration of L-deprenyl may enhance the abnormal involuntary movements and therefore in patients with dyskinesia its administration should be avoided. Bromocriptine as an adjuvant agent seems to be sufficient in patients with rather regular oscillation between dyskinetic and akinetic periods. Tiapride is able to reduce disturbing peak dose dyskinesia or paroxysmal dyskinesia. An hourly observation system enables the more correct determination of the type of on-off phenomena and the selection of the medicament of choice.

Patients and Methods

In order to get better information about the diurnal fluctuations of the motor performance of our patients we have introduced a new, short and simple rating scale, called the motor performance scale (MPS). The items of MPS are presented in Tables 1-4.

As standing up and gait may be equally disturbed by either hypokinesia or dyskinesia, both possibilities were taken into account. For example gait=4 AIM=0 means that the patient is severely akinetic. Gait=4 AIM=4 means that the patient is unable to walk without assistance owing to abnormal involuntary movements. Making a diagram (Fig. 1): the area under the *dotted line* and above the x-axis shows the values of standing up + gait, what we call "mobility". The whole area between the *continuous lines* on both sides of the x-axis is the "total disability". Both areas can be expressed numerically by a selective mathematical programme of area integration. (First figure = mobility, second figure = total disability.) The patients taking part in the trial showed various "on-off" symptoms. According to the type of their most characteristic diurnal oscillation the patients were divided into three therapeutic groups.

Table 1. Motor performance scale: assessment of tremor

0	No tremor
1	Minimal, intermittent tremor at rest; occasionally activated when walking
2	Slight intermittent tremor at rest, amplitude less than 2.5 cm
3	Almost constant tremor at rest, amplitude less than 10 cm
4	Constant tremor at rest, with amplitude greater than 10 cm

Table 2. Motor performance scale: assessment of standing up (from a chair)

0	Normal
1	Slightly slower; slightly uncoordinated owing to abnormal involuntary movements (AIM)
2	Moderately slower or uncoordinated owing to AIM; support by hands is needed
3	Markedly slower, or aggravated by AIM; falls back, but after repeated attempts stands up without assistance
4	Unable to get up without assistance

AIM 0-2 = standing up 0
AIM 3 = standing up 0-3
AIM 4 = standing up 4

Table 3. Motor performance scale: assessment of gait

0	Normal; in spite of AIM gait remains undisturbed, including starting, turning and stopping
1	Slightly slower gait, shorter steps (30-45 cm); no shuffling at start, nor when walking or turning; gait slightly uncoordinated owing to AIM; direction and length of steps variable
2	Moderately slower gait, shorter steps (15-30 cm); occasional shuffling; on starting and on turning, feet adherer to floor; gait moderately uncoordinated and unstable owing to AIM; no need of support
3	Very slow, shuffling gait; steps are shorter than 15 cm; starting and turning severely disturbed; very uncoordinated and unstable gait owing to AIM; occasional assistance needed
4	Unable to walk without assistance, severe akinesia or AIM; bedridden

AIM 0-2 = gait 0
AIM 3 = gait 0-3
AIM 4 = gait 4

Table 4. Motor performance scale: assessment of abnormal involuntary movements

0	No AIM
1	Slight, intermittent, occasionally activated when moving
2	Constant AIM which do not disturb standing up, gait or manual activity
3	AIM causing evident disturbance of standing up, gait or manual activity
4	Severe, constant AIM; needs assistance with standing up and walking; manual activity badly disturbed

AIM 0-2	= standing up	0
	= gait	0
AIM 3	= standing up	0-3
	= gait	0-3
AIM 4	= standing up	4
	= gait	4

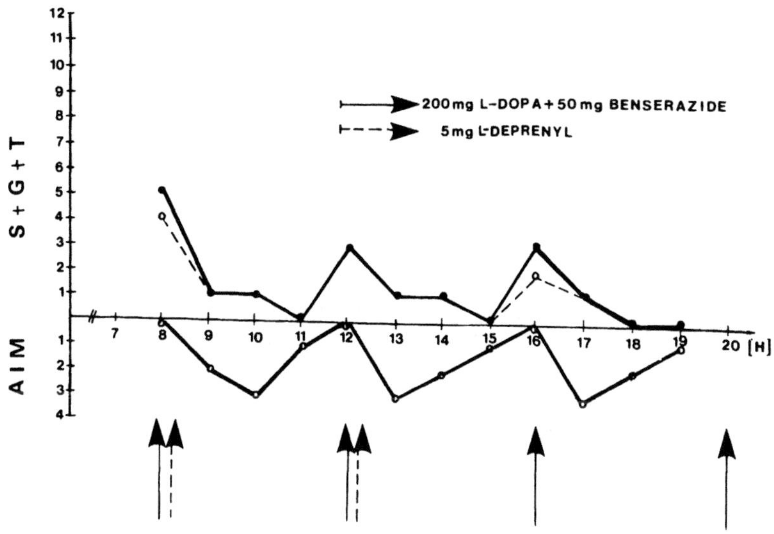

Fig. 1. Daily diagram based on the motor performance scale; for details see text

Results

In group I 11 patients were treated with 2x5 mg L-deprenyl daily, without any other changes in the earlier therapy. Duration of illness was 3-16 years, the duration of on-off phenomena, including the earliest signs, such as slight wearing off or freezing, was 1-7 years (mean: 3.9 years). All of them had end-of-dose failure, but only two showed regular dyskinesias after the intake of a single dose of Madopar. Mean dose of L-dopa + peripheral decarboxylase inhibitor (PDI) was 544+136 mg/day.

After 14-21 days the MPS values showed a significant improvement. Six patients improved considerably, but neither of the two patients with hyperkinesias; in these patients the off periods became somewhat shorter, but stronger hyperkinesias developed.

In group II ten patients were treated with *bromocriptine* as an adjunct to earlier L-dopa + benserazide therapy. The duration of illness was 7-18 years and the onset of on-off phenomena 1-5 years (mean 2.75 years). Eight patients showed end-of-dose failure with a higher incidence of abnormal involuntary movements in the on phase. Two patients had "yo-yo" symptoms. Mean dose of L-dopa + PDI was 600+150 mg/day.

We have used bromocriptine as an adjuvant, starting with 1.25 mg/day. We increased the dose very cautiously to 7.5-20 mg/day, normally in 2-3 weeks. On average it allowed a 17% dose reduction of L-dopa + PDI. The improvement of eight patients was very satisfactory, as their off periods became shorter, and in the on periods only mild abnormal involuntary movements developed. All the improved patients belonged to the subgroup with end-of-dose failure. Two patients with non-dose-related on-off phenomena did not show any amelioration. Figure 2 shows the improvement of a patient with dose-related on-off disturbances after the introduction of adjuvant bromocriptine therapy.

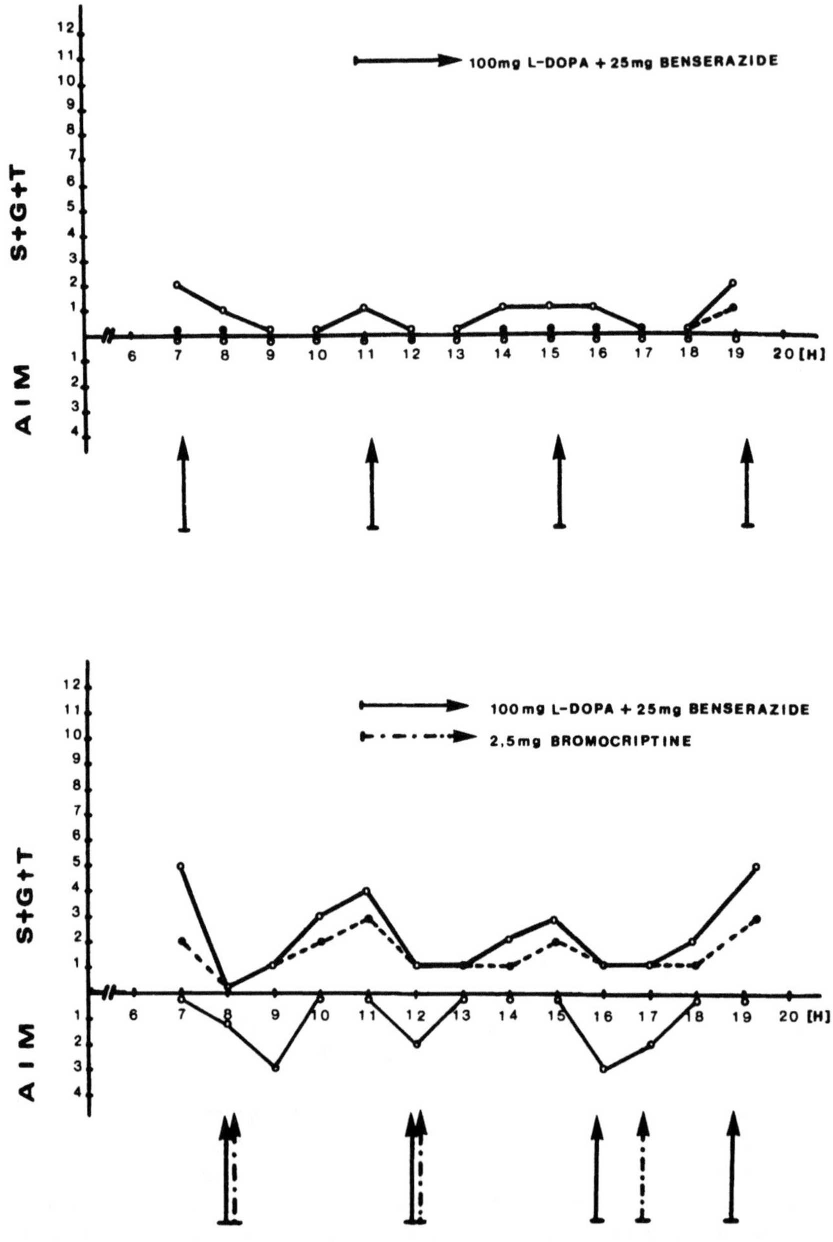

Fig. 2. Improvement of a patient with dose-related on-off symptoms after introducing supplementary bromocriptine therapy. Abnormal involuntary movements disappeared, the daily motor performance improved

Table 5. On-off phenomena: effects of L-deprenyl, bromocriptine and tiapride

Type[a]	L-Deprenyl $N = 11$ End-of-dose failure, 9 wearing off	Bromocriptine $N = 10$ 8 End-of-dose failure, 2 "yo-yo" symptoms	Tiapride $N = 7$ End-of-dose failure, peak dose dyskinesia
MPS			
Mobility	$10.0 \pm 4.0 \rightarrow 6.36 \pm 2.3^{d}$	$21.5 \pm 3.8 \rightarrow 10.9 \pm 2.4^{e}$	$18.0 \pm 4.14 \rightarrow 11.9 \pm 3.92^{b}$
Total disability	$14.7 \pm 3.3 \rightarrow 12.8 \pm 3.6^{d}$	$39.9 \pm 3.8 \rightarrow 21.7 \pm 2.95^{e}$	$33.0 \pm 8.1 \rightarrow 16.4 \pm 3.1^{c}$
NWUD	$69.5 \pm 5.0 \rightarrow 75.6 \pm 3.74^{c}$	$67.6 \pm 3.93 \rightarrow 77.4 \pm 4.92^{f}$	$78.2 \pm 3.48 \rightarrow 86.3 \pm 2.97^{g}$
Improved	6	8	7
Unchanged	5	2 (Yo-yo)	0

[a] MPS motor performance scale; NWUD Northwestern University disability scale.
[b] Not significant; [c] $P < 0.05$; [d] $P < 0.02$; [e] $P < 0.01$; [f] $P < 0.005$; [g] $P < 0.001$.

In group III the patients have been characterized by end-of-dose failure. The duration of illness was 2-12 years (mean: 8.4 years), the onset of on-off phenomena was between 1-5 years (mean: 2.3 years). The mean dose of L-dopa + PDI was 514+128.5 mg/day. All of them had regular, disturbing peak dose dyskinesia, which made their gait and balance unstable. In a female patient we observed paroxysmal dyskinesia; in this period she had severe abnormal involuntary movements and became bedridden. In this group tiapride was administered. The dose had to be titrated very carefully, as in some patients higher doses (100 mg orally) caused deep and prolonged akinesia. The average dose in this group was only 75-150 mg/day, that is three doses of one-quarter or one-half tablet. In some patients a slight increase of L-dopa + PDI was made, but on average only by 8.3%. With this regimen the values of mobility did not change significantly, while total disability values, more related to dyskinesia, were reduced. Our results for the three groups are shown in Table 5.

Discussion

The results of our therapeutic trial can be summarized as follows.

L-Deprenyl

L-deprenyl is useful in patients with wearing-off symptoms, mostly without dyskinesia. In these patients L-deprenyl might make the postsynaptic action of dopamine longer owing to its MAO-B and reuptake inhibiting effect. Dyskinetic periods became longer, which seems to limit the use of this drug.

Bromocriptine

Bromocriptine, as an adjuvant agent, is very helpful in patients with end-of-dose failure. Both dyskinetic and akinetic periods became shorter, owing either to a direct postsynaptic or presynaptic effect of the drug. It is worth mentioning that bromocriptine had no beneficial effect in patients with "yo-yo" symptoms.

Tiapride

Tiapride is useful, if peak dose dyskinesia or paroxysmal dyskinesia are present. Its dose must be titrated very carefully as it can make off periods deeper and longer. We have to emphasize that the use of hourly observation, done by trained nurses and medical students under the control of a physician, is a great help to define the type of on-off phenomena and to select the medicament of choice.

References

Birkmayer W (1978) Long-term treatment with L-deprenyl. J Neural Transm 43: 239-244

Csanda E, Fornádi F (1982) Experience clinique du tiapride dans le traitement des syndromes extrapyramidaux. Sem Hop Paris 58: 2647-2650

Csanda E, Antal J, Antony M, Csanaky A (1978) Experiences with L-deprenyl in parkinsonism. J Neural Transm 43: 263-269

Csanda E, Tárczy M, Takáts A, Mogvorós I, Köves Á, Katona G (to be published) L-deprenyl in the treatment of Parkinson's disease. J Neural Transm

Lees AJ, Lander CM, Stern GM (1979) Tiapride in the levodopa induced involuntary movements. J Neurol Neurosurg Psychiatry 42: 380-383

Le Witt PA, Calne DB (1981) Recent advances in the treatment of Parkinson's disease. The role of bromocriptine. J Neurol Transm 51: 175-184

Lhermitte F, Signoret JL, Agid Y (1977) Etude des effets d'une molécule originale, le tiapride, dans le traitement des mouvements anormaux d'origine extrapyramidale. Sem Hop Paris 53: 9-15

Marsden CD, Parkes JD (1976) "On-off" effects in patients with Parkinson's disease on chronic levodopa therapy. Lancet 1: 292-296

Tárczy M (1983) New rating scale for the evaluation of "on-off" phenomena. Ideggyogy Szle 36: 363-371

Teychenne PF, Bergsrud D, Racy A, Vern B (1981) Low dose bromocriptine therapy in Parkinson's disease. Res Clin Forums 3: 37-53

Yahr MD (1978) Overview of present day treatment of parkinson's disease. J Neural Transm 43: 227-238

Target and Additive Drugs in the Treatment of Parkinson's Disease: A Pilot Study

W. Birkmayer and P. Riederer

Introduction

The immediate cause of Parkinson's disease is a deficit of various neurotransmitters owing to progressive degeneration of the nigrostriatal dopaminergic system and other brain structures like the locus coeruleus which is mainly involved in noradrenergic neurotransmission, while reduced serotonergic activity and loss of GABAergic function might be a secondary consequence of such disturbances (Ehringer and Hornykiewicz 1960; Bernheimer et al. 1961). Therefore, optimal treatment of parkinsonism should be directed primarily to antagonize the degenerative process. This, however, will only be possible when the cause of the disease process has been found. Today, administration of drugs balancing the disturbed dopaminergic neuronal activity is the treatment of choice. Such drugs are target drugs, while others acting on various other sites of neurotransmission (i.e. cholinergic, serotonergic, GABAergic systems) and influencing the dopaminergic activity indirectly are additive drugs.

Target Drugs in Parkinsonian Therapy

An optimal treatment of Parkinson's disease is to substitute a deficiency of dopamine, noradrenaline and serotonin by the precursor amino acids *l*-dopa, 3,4-dihydroxyphenylserine (DOPS) and *l*-tryptophan or 5-hydroxytryptophan, respectively. To improve the supply of *l*-dopa in the brain and to reduce peripheral dopa accumulation and subsequent peripheral side effects, introduction of a combined medication of *l*-dopa with benserazide (62.5 mg t.i.d. - 250 mg b.i.d.) and carbidopa (25-250 mg) (benserazide and carbidopa being peripherally acting decarboxylase inhibitors) led to a substantial improvement of this therapeutic strategy (Birkmayer and Mentasti 1967; Birkmayer and Riederer 1980).

Another method of target medication is the stimulation of postsynaptic dopaminergic receptors and the subsequent influence of other neurotransmitter receptor systems. Bromocriptine, lisuride and pergolide (all three stimulate dopaminergic receptors) are antagonists of noradrenergic β-receptors and agonists of serotonergic receptors, although their actions are quantitatively different (Loew 1981). Unfortunately, these so called dopaminergic agonists: (a) are not selective with respect to the dopaminergic system; and (b) stimulate both pre- and postsynaptic receptor units, causing a further depression of presynaptic dopamine synthesis via feedback inhibition (Loew 1981). This theoretical assumption, however, might be of clinical significance, at least in patients on short-term treatment with dopaminergic agonists or in patients for whom treatment with dopaminergic agonists has been started. In such patients an increased disability has been noted during the first few hours after onset of therapy, while later the beneficial effects on disability seem to be correlated with a preferred stimulation of postsynaptic receptors. Dopaminergic agonists are given preferentially in advanced parkinsonism (Birkmayer and Riederer 1980).

Another type of target drugs are selective monoamine oxidase (MAO) inhibitors, specifically affecting the dopaminergic neuron. So far, only two substances have been shown to be of clinical importance as selective MAO-B inhibitors (the substrates dopamine or serotonin). These are deprenyl and (+)-tranylcypromine. During long-term treatment both substances are selective only at low doses (1 mg per 10 kg body weight for deprenyl and 0.6 mg per 10 kg body weight for (+)-tranylcypromine (Birkmayer and Riederer 1980).

Additive Drugs in Parkinsonian Therapy

There is general agreement that the efficacy of target drugs drops with the progress of Parkinson's disease. Owing to severe denervation intra- and interneuronal imbalances occur. As a consequence adverse reactions and side effects are more frequent and less tolerated in the advanced stage of Parkinson's disease. High doses of target drugs might have deleterious effects on imbalanced dopaminergic neurons. Thus, antiparkinson drugs should be administered in low doses if possible and additives potentiating the beneficial effects of target drugs should be given.

Anticholinergics are widely used in addition to target drugs. A preliminary clinical trial (between 1974 and 1976) with prodipine in 161 patients with Parkinson's disease showed excellent to moderate improvement of resting tremor in 60 of the patients, while medication had to be discontinued in 98 patients owing to gastrointestinal side effects and hypotension.

Budipine, a drug with weak anticholinergic properties, stimulates the catecholaminergic system with enhancement in vigilance and motor activity. In animals pretreated with reserpine, dopamine, noradrenaline and 5HT levels increase in the brain after budipine treatment while MAO activity remains unchanged. As the drug does not displace spiroperidol from receptor sites an interference of budipine with so called D2 receptors is unlikely (Menge and Brand 1982).

In the last 2 years we have administered budipine (10-25 mg t.i.d.) orally in addition to low doses of combined *l*-dopa treatment. A remarkable antitremor effect could be noted. More recently, intravenous budipine (10-20 mg) has been administered to patients with akinetic crises. Beneficial effects have been observed in 51 of 64 patients. Improvement of disability scores (Birkmayer and Neumayer 1972) gave evidence of a dopaminergic action although in some patients side effects were more frequent than with prodipine.

Although the biochemical profile of budipine is not fully understood the increase in dopamine and noreadrenaline concentration observed in experimental studies is in agreement with an improvement of motor function in patients with akinetic crisis. Moreover, a facilitory action of noradrenaline on dopaminergic transmission might be of significance. In this respect, budipine is of therapeutic value. Furthermore, the beneficial antitremor efficacy reflects different pharmacological properties from *l*-dopa or most of the dopaminergic agonists which in most cases have only weak antitremor effect.

Conclusion

Additives like budipine resemble drugs which improve the therapeutic effect of target drugs both by functional optimization and by having additional pharmacological properties directed to improve neurotransmission in patients with multineuronal disturbance.

References

Bernheimer H, Birkmayer W, Hornykiewicz O (1961) Verteilung des 5-Hydroxytryptamins (Serotonin) im Gehirn des Menschen und sein Verhalten bei Patienten mit Parkinson-Syndrom. Wien Klin Wochenschr 73: 1056

Birkmayer W, Mentasti M (1967) Weitere experimentelle Untersuchungen über den Katecholaminstoffwechsel bei extrapyramidalen Erkrankungen (Parkinson- und Choreasyndrom). Arch Psychiatr Nervenkr 210: 29-35

Birkmayer W, Neumayer E (1972) Die moderne medikamentöse Behandlung des Parkinsonismus. Z Neurol 202: 257

Birkmayer W, Riederer P (1980) Die Parkinson-Krankheit. Springer, Vienna New York

Ehringer H, Hornykiewicz O (1960) Verteilung von Noradrenalin und Dopamin im Gehirn des Menschen und ihr Verhalten bei Erkrankungen des extrapyramidalen Systems. Klin Wochenschr 38: 1236-1239

Loew DM (1981) Die Wirkung von Mutterkorn-Derivaten auf das Zentralnervensystem: Synaptische Übertragungsmechanismen. In: Platt D (ed) Funktionsstörungen des Gehirns im Alter. Schattauer, Stuttgart New York, pp 83-101

Menge HG, Brand U (1982) Pharmacological summary of budipine, a new 4,4-diphenylpiperidine derivative used in treating Parkinsons's disease. Arzneimittelforsch 32/1: 85

Pharmacologic Bases of Antiparkinsonian Therapy

H.G. Menge and U. Brand

The pathologic mechanism underlying Parkinson's disease, dopamine deficiency in the striatum (Ehringer and Hornykiewicz 1960), may be therapeutically redressed in several ways: by administering anticholinergics to reduce the cholinergic surplus, by substituting for the lack of transmitter by giving levodopa, by direct stimulation of dopaminergic receptors, or by inhibition of dopamine reuptake. Whether a pharmacotherapeutic measure is successful will ultimately depend on the stage to which the disease has progressed, the length of treatment provided, and the level of dosing applied. It is generally known, for example, that dopaminergic receptors themselves are changed both in density and responsiveness as a result of dopamine depletion and pharmacotherapy (Jellinger et al. 1982; Rinne et al. 1981). There is a clear correlation between progressive pathologic events and the type of treatment provided. In practice, the combination of different therapies has proved highly useful. It appears that combination therapy helps to prevent an excessive, one-sided influence on the transmitter systems implicated in the disease process (Fischer 1980).

Budipine is a new antiparkinsonian drug. It has been developed from a series of diphenylpiperidine derivatives after extensive pharmacologic screening and intentisve subsequent studies. Another derivative, the isopropyl compound of diphenylpiperidine, turned out to have an inadequate effect when administered orally. In budipine, the isopropyl group has been replaced by a tertiary butyl group (Fig. 1), which renders its metabolic degradation at the basic nitrogen atom much more difficult (Schäfer et al. 1984).

The following describes the pharmacologic profile of budipine. Since animals are not susceptible to parkinsonism, the pharmacologist must rely on suitable "models" for simulating the symptoms of Parkinson's disease. Comparative studies on activity will have to be run with drugs known to be effective in Parkinson's disease.

Tremor Models

By stimulating muscarinic receptors with oxotremorine, fine tremor can be induced besides autonomic effects such as salivation, hypothermia and lacrimation (Cho et al. 1962; George et al. 1962). Compounds acting on parkinsonian tremor dose-dependently inhibit both tremor and autonomic symptoms on preventive administration (Nose and Kogunia et al. 1970; Setley and van Woert et al. 1973). Tremor induced by physostigmine is also of the muscarinic type (Gothoni et al. 1983; Seth and van Woert 1973). It

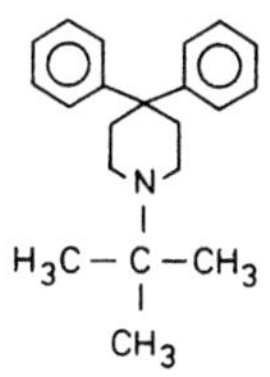

Fig. 1. Budipine: 1-*tert*-butyl-4,4-diphenylpiperidine

is not caused by direct receptor stimulation, as in the case of oxotremorine, but is due to accumulation of endogenous acetylcholine from inhibition of acetylcholinesterase. The third tremor model, which involves the compound LON-954, *N*-carbamoyl-2 (2,6-dichlorophenly) acetamidine hydrochloride, is not of the muscarinic type but, according to Coward et al. (1977) and Brand and Menge (1980), is of a dopaminergic character and based on an artificially induced dopamine deficit. For tremor inhibition by budipine and conventional antiparkinsonian drugs, see Table 1.

Akinesia Models

Haloperidol, a classic neuroleptic, blocks mainly dopaminergic D2 receptors (Seeman et al. 1975). Perphenazine, by contrast has more α_1-adrenoceptor blocking properties (Janssen 1965). In small animals, the effect is manifest as catalepsy and can be quantitatively assessed in the mouse (two parallel metal wires: Julou et al. 1967) and rat (7-cm block; Stille 1971). With this experimental design, the antagonistic effect of budipine and other antiparkinsonian drugs can be easily demonstrated, both preventively and curatively (Table 2).

Rigidity Models

An akineto-rigid syndrome can be elicited by depletion of noradrenaline, dopamine, and also serotonine due to reserpine or tetrabenazine, inducing catalepsy as after administration of a neuroleptic, plus rigidity of the skeletal muscles (Carlsson 1975; Reches et al. 1983; Jurna 1968). The methodology used here is the same as for the akinesia model. The results with budipine and the other antiparkinsonian drugs are summarized in Table 3.

Budipine proved effective in all of these models. The other four drugs were effective to the extent that their particular mechanism of action, as simulated by the respective model, was implicated. This explains why there may be no effect at all or a different effect from that of budipine, both qualitatively and quantitatively. With the studies conducted to date, we have not been able to explain fully the molecular mechanism responsible for this broad spectrum of activity. Highly specific studies will have to be carried out.

Table 1. Effects of different antiparkinsonian drugs on various tremor models

Substance/model	ED_{50} (mg/kg p.o.)				
	Budipine	L-Dopa + decarboxylase inhibitor	Bromocriptine	Amantadine	Biperiden
Oxotremorine					
Mouse	36	Synergism	No effect	Synergism	20
Rat	22	Synergism	No effect	Synergism	30
Physostigmine					
Mouse	12	Not evaluated	Not evaluated	80	53
LON-954					
Mouse	15	70	No effect	60	10

Table 2. Effect of different antiparkinsonian drugs on two akinesia models

Substance/model	ED_{50} (mg/kg p.o.)				
	Budipine	L-Dopa + decarboxylase inhibitor	Bromocriptine	Amantadine	Biperiden
Haloperidol					
Mouse					
Preventive	15	No effect	Weak effect	60	3
Curative	20	250	Weak effect	50	4.5
Rat					
Preventive	50	No effect	Not evaluated	50	10
Curative	24	Weak effect	No effect	25	12
Perphenazine					
Mouse					
Preventive	13	70	Not evaluated	26	1
Curative	10	Weak effect	Not evaluated	45	1
Rat					
Preventive	13	180	Not evaluated	14	4.5
Curative	25	No effect	Not evaluated	16	3.6

Table 3. Effects of different antiparkinsonian drugs on various rigidity models

Substance/model	ED_{50} (mg/kg p.o.)				
	Budipine	L-Dopa + decarboxylase inhibitor	Bromocriptine	Amantadine	Biperiden
Reserpine					
Mouse					
Preventive	Weak effect	125	8.2	100	Weak effect
Curative	35	65	28	250	Weak effect
Rat					
Preventive	45	57	20	5	Weak effect
Curative	9	60	8	26	Weak effect
Tetrabenazine					
Mouse					
Preventive	50	100	3	Weak effect	No effect
Curative	34	85	Weak effect	75	20
Rat					
Preventive	2.6	17	10	19	3
Curative	14	70	8	8	6

Anticholinergic Properties

The classical design (Magnus 1904) on isolated guinea pig ileum, yields a pA_2 value of 6.68 for budipine. Analogous values for atropine and biperiden are 8.99 and 8.13. Amantadine, L-dopa, and bromocriptine have no anticholinergic activity. Further results in peripheral, cholinergically innervated organs (acetylcholine-induced contractions in rat stomach; acetylcholine-induced contractions in rat urinary bladder; carbachol-induced negative inotropism in rat atrium; carbachol-induced contractions in guinea pig trachea) are listed in Table 4 for budipine, atropine, and biperiden as relative effects in terms of budipine = 1.

Binding studies with cortex and myocardium homogenates also reveal differences in affinity to acetylcholine receptors for budipine and biperiden: pA_2 values in the cortex (ligand propylbenizilylcholine ^3H) are 6.83 (budipine) and 8.16 (biperiden); in the myocardium (lingand *N*-methylscopolamine ^3H) pA_2 values are 6.53 (budipine) and 7.42 (biperiden) (Birdsall 1982). These studies are in full agreement with the previous results obtained in isolated organs (Table 4).

Table 4. Anticholinergic effects of biperiden and atropine relative to budipine (n = 1)

On basis of receptor affinities (pA_2 values)

Receptor	Biperiden	Atropine
Propylbenzilylcholine ^3H (cortex)	21.4	100
N-Methylscopolamine ^3H (myocardium)	7.8	100

On basis of peripheral effects in vivo (ED_{50} values)

Experiment	Biperiden	Atropine
Carbachol-salivation rat p.o.	100	1500
Oxotremorine-salivation rat p.o.	38	100
Mydriasis mouse s.c.	130	1300
ACh-induced blood pressure decrease rat i.v.	46	541

On basis of peripheral effects in vitro (pA_2 values)

Organ	Biperiden	Atropine
Ileum	28	204
Trachea	17	1100
Stomach	98	229
Bladder	52	427
Heart atrium	1.4	174

In in vivo studies (mouse) with subcutaneous administration of compound, an enlargement of the pupilla was observed after budipine administration (ED_{50} ~13 mg/kg). Corresponding ED_{50} values for atropine and biperiden are 0.01 and 0.1 mg/kg, respectively. Also in vivo, a similar activity pattern results for inhibition of carbacholstimulated salivation in the anesthetized rat. ED_{50} values are 75 mg/kg for budipine, 0.05 mg/kg for atropine, and 0.74 mg/kg for biperiden, all with oral administration (Table 4). Gastrointestinal peristalsis of the mouse is inhibited up to 30% by budipine at 50 mg/kg s.c. The anticholinergic effect of budipine observed in cardiovascular studies in the rat and cat were not demonstrable on heart rate, which is regularly lowered under budipine. An acetylcholine-induced fall of blood pressure in the rat is attenuated by budipine with an ED_{30} of 6.5 mg/kg i.v. The ED_{30} for biperiden is 0.14 mg/kg i.v., and 0.012 mg/kg i.v. for atropine.

Dopaminergic Effects

These studies were concerned mainly with the specific influence on D2 receptors, which are of particular importance in Parkinson's disease. In an in vitro study, Czlonkowski (personal communication, 1978) showed that spiroperidol ^{3}H is not displaced by budipine in homogenates of rat striatum. In the intact animal, budipine was not observed to cause stereotypies of the kind observed after apomorphine administration in rat and mouse. We merely found a slight potentiation of subacute apomorphine. Also, a subacute L-dopa dose was found to be potentiated by budipine (at 10-50 mg/kg p.o.) in the mouse. In rats with unilateral lesions of the nigrostriatal pathway, budipine did not induce turning behaviour of the kind produced by apomorphine, not even when sensitizing the postsynaptic dopaminergic receptor (D2) by 3-day pretreatment with haloperidol. There is an increase in the binding capacity in the corpus striatum after chronic administration of neuroleptics (Burt et al. 1977 and others). This increase in the binding capacity after repeated administration of haloperidol can be partially antagonized by subsequent 5-day administration of L-dopa. Budipine influence on D2 receptors after multiple administration was studied with the method of List and Seeman (1979) by H.H. Hennies (1983, unpublished work). Hennies found by way of determining D2 receptor density in the striatum (Sundermann and Wooten 1980) that, on twice daily administration of budipine for 14 days at 10 mg/kg s.c. in addition to chronic haloperidol administration, the increase in binding capacity can be fully or partially antagonized. Effects are also demonstrable on peripheral dopaminergic receptors which are manifest in a steeper rise of blood pressure after pretreatment with budipine (Menge and Brand 1982).

Noradrenergic Effects

During the first tests, we observed a marked central stimulation. On closer analysis, it appeared to distinguish itself from the classic stimulants: there was no increase in salivation, no increase in temperature, no group toxicity. Also, appetite was not reduced in the rat. In discrimination tests, rats trained to react to amphetamine clearly differentiated between that stimulant and budipine. Also, there was no evidence of a cardiovascular reaction as after administration of stimulants. Blood pressure rises briefly and slightly in the rat, cat and dog; bradycardia occurs as previously stated.

In isolated rat vas deferens, budipine causes contractions at high concentrations only (10^{-4} mol/l). Noradrenaline and tyramine effects are enhanced from 1×10^{-6} mol/l and 3×10^{-6} mol/l, respectively. There is no evidence of an intrinsic effect of budipine in chemically denervated organs (with 6-hydroxydopamine). Following reserpine-induced release of noradrenaline from storage granule, budipine is able to elicit contractions in the isolated vas deferens at 10^{-4} mol/l. No α_2-agonistic or α_2-antagonistic effects have been demonstrated in electrically stimulated guinea pig ileum strips compared with clonidine or yohimbine, respectively. Monoamine oxidase (MAO) inhibitory effects can be excluded.

Serotonergic Effects

In in vivo studies in rats, head twitches induced by the serotonergic receptor agonist quipazine were enhanced by budipine from 5 mg/kg p.o. Further, pretreatment with budipine (12.5-50 mg/kg p.o.) inhibited the effect of the 5-HT releaser chloramphetamine (Menge and Brand 1982). Fenfluramine-induced 5-HT depletion in the brain was prevented partially, but stimulation of the hindlimb flexor reflex in spinal rats by LSD or fenfluramine was not antagonized by budipine (Przegalinski et al. 1982). In isolated human blood platelets, budipine inhibits both reuptake and release of ^{14}C-5-HT with an IC_{50} of 9×10^{-5} and 7.4×10^{-4} mol/l, respectively. For amantadine, the values are 1.3×10^{-4} mol/l for reuptake inhibition and 4×10^{-3} mol/l for release (Eltze 1980).

Influence on Content and Turnover of Biogenic Amines

The influence of budipine as demonstrated in the different models may be also reflected in changes in the content and turnover of biogenic amines. Following single administration of 50 mg/kg budipine, no significant changes were observed in the content of noradrenaline, dopamine, and serotonin in whole brain. Likewise, neither noradrenaline nor dopamine turnover was altered after a single dose of 50 mg/kg budipine p.o. and 8 mg/kg budipine p.o. on 5 consecutive days.

Discussion

Budipine has a pharmacologic profile that does not readily permit one to classify the compound with any one of the known antiparkinsonian agents. In the normal animal involved in pharmacologic and toxicologic screening studies with acute and chronic administration, no other effect was observed except central stimulation, which clearly distinguishes itself from that of other central stimulants. The situation is different in the model experiments, where even low doses of budipine either abolish or improve a severe pathologic condition (see also the chapter by Przuntek and Stasch). A similar picture emerged in the clinical studies. The healthy subject, under the influence of single doses in excess of those relevant in Parkinson's disease, exhibits primarily symptoms indicating central stimulation. Side effects or concomitant effects, such as occur typically with anticholinergic therapy, are hardly seen, especially after high doses of budipine. This is consistent with the weak anticholinergic activity exhibited in comparison with atropine and biperiden, a finding also made in the pharmacologic studies previously referred to. We do not think the therapeutic effect of budipine is exclusively due to this anticholinergic component.

Considering the likelihood of interactions as a result of the close links between the cholinergic, dopaminergic (among others Anden et al. 1966; Sethy and van Woert 1974), and possibly other systems as well, strictly one-sided monotherapy appears to

Table 5. Transmitter systems involved by budipine; results refer to the specific system

Noradrenergic effects	
Behaviour:	central stimulation, potentiation of amphetamine
Models:	antagonism against reserpine
Autonomic:	potentiation of noradrenaline in isolated vas deferens, weak elevation of blood pressure
Dopaminergic effects	
Behaviour:	potentiation of L-dopa and apomorphine
Models:	antagonism against haloperidol, reserpine, tetrabenazine, and LON-954
Autonomic:	potentiation of dopamine blood pressure effect
Receptors:	antagonism against augmentation of receptors after haloperidol administration
Serotonergic effects	
Behaviour:	potentiation of quipazine, antagonism against cloramphetamine
Autonomic:	contraction of isolated stomach strips
Blood platelets:	inhibition of reuptake and 5-HT release
Cholinergic effects	
Models:	antagonism against oxotremorine and physostigmine
Autonomic:	mydriasis, antagonism against acetylcholine in isolated organs
Receptors:	affinity to cortex and myocardial receptors

be less promising of success than the influence exerted by combination therapy on the different transmitter systems.

The pharmacologic and clinical results suggest an influence by budipine on functions that are an integral part of the cholinergic, dopaminergic, and serotonergic system. For a survey, see Table 5. It is very difficult to elucidate the molecular pharmacology behind the effect on each of these systems. As for the dopaminergic system, a direct antagonistic effect such as observed after administration of apomorphine, piribedil, bromocriptine, and other drugs can be ruled out. The results obtained by H.H. Hennies (1983, unpublished work), i.e., antagonism by budipine of the increase in receptor density after haloperidol administration, are certainly relevant since we are dealing here, as in Parkinson's disease, with a correction of pathologic events.

The situation ist still unclear as regards the noradrenergic effect. Reuptake inhibition, as in the case of the classic antidepressants, and MAO inhibition are excluded. Direct stimulation of noradrenergic receptors is not very likely, while a release from

Table 6. Effects of various therapeutic agents with different modes of action on parkinsonian models. Values are ca. ED_{50} in mg/kg p.o. at the time of peak effect. In the tremor models, the agents were given before the tool substance (preventively), in the rigidity and akinesia models afterwards (curatively), i.e., in the presence of the symptom

Model	Tool substance	Influence on transmitter	Budipine	L-Dopa	Bromocriptine	Amantadine	Biperiden
Tremor	Physostigmine	Cholinergic					
	Mouse		25	75	30	80[a]	100-150
	Oxotremorine						
	Mouse		36	Synergism	No effect	Synergism	20
	LON-954	Dopamine					
	Mouse	deficit	15	70	No effect	60	10
Rigidity	Reserpine						
	Mouse	Noradrenaline	35	65	30	45	Weak effect
	Curative	Dopamine					
		Serotonin					
	Rat	Depletion	9	60	8	26	Weak effect
	Curative						
Akinesia	Haloperidol						
	Mouse	Dopamine	20	250	Weak effect	50	4.5
	Curative	Receptor antagonist[a]	24				
	Rat			Weak effect	No effect	25	12
	Curative						

[a] s.c. application.

reserpine-resistant storage granule has in effect been demonstrated. The central stimulatory component on which these phenomena rest can easily be associated with an anti-akinetic property.

In order to extrapolate the pharmacologic behaviour of budipine to the therapeutic effect in humans, we must reintegrate each individual effect and obtain a global impression. Table 6 summarizes the findings made in the tremor, rigidity, and akinesia models. This already examplifies the problems faced in predicting the clinical outcome on the basis of pharmacologic performance. Using L-dopa plus decarboxylase inhibitor, effects are achieved in animal experiments in some cases with high doses only. The direct receptor stimulant bromocriptine acts against reserpine, but not against LON-954 or haliperidol. On the other hand, tremor of any genesis and haloperidol-induced receptor blockade are easily antagonized by biperiden. Yet, biperiden is not effective in the rigidity model against reserpine, where budipine in turn is highly effective. The pharmacology of amantadine, which is successfully applied clinically, is not easily interpreted. Relatively high doses are required to achieve an effect. Here, too, the mechanism of action is still subject to clarification (Menon et al. 1984). It illustrates, however, that there are therapeutic roads still left to be travelled that have hitherto defied all therapeutic concepts.

The tabulated survey shows that from a pharmacologic point of view budipine is effecitve in all models by acting on different transmitter systems. This is not true of the other antiparkinsonian agents. Therapy of Parkinson's disease is best approached from different angles. By influencing several systems, the broad spectrum of activity of budipine may turn out to be a particular asset.

References

Anden NE, Dahlström A, Fuxe K, Larsson K (1966) Functional role of the nigro-neustriatal dopamine neurones. Acta Pharmacol Toxicon (Copenh) 24: 263-274

Birdsall NJM (1982) National Institute for Medical Res. London (unpublished results, data on file)

Brand U, Menge HG (1980) Antagonismus gegen den Tremor nach N-Carbamoyl-2(2.6-dichlorphenyl-acetamidin-HCl (LON-954) und Oxotremorin. Eine vergleichende Studie zum Wirkungsmechanismus. Arzneimittelforsch 30: 1242-1243

Burt DR, Creese J, Snyder SH (1977) Antischizophrenic drugs: chronic treatment elevates dopamine receptor binding in brain. Science 196: 326-328

Carlsson A (1975) Monoamine-depleting drugs. Pharmacol Ther B 1: 393-400

Cho AK, Haslett WL, Jenden DJ (1962) The peripheral actions of oxotremorine, a metabolite of tremorine. J Pharmacol Exp Ther 138: 249

Coward DM, Doggett NS, Thomas JE (1977) Central transmitter involvement in LON-954-induced tremorgenesis. Neuropharmacology 16: 479-484

Ehringer H, Hornykiewicz O (1960) Verteilung von Noradrenalin und Dopamin (3-Hydroxytyramin) im Gehirn des Menschen und ihre Verhalten bei Erkrankungen des extrapyramidalen Systems. Klin Wochenschr 38: 1236

Eltze M (1980) The effect of prodipine and budipine on ^{14}C-5-Hydroxytryptamine uptake and release by human blood platelets. Arzneimittelforsch 30: 1129-1134

Fischer PA (ed) (1980) Parkinson-Syndrom: Kombinations- und Begleit-Therapien. Schattauer, Stuttgart

George R, Haslett WL, Jenden DJ (1962) The central action of a metabolite of tremorine. Life Sci 1: 361

Gothóni, P. Lehtinen M, Fincke M (1983) Drugs for Parkinson's disease reduce tremor induced by physostigmine. Naunyn+Schmiedebergs Arch Pharmacol 323: 205-210

Janssen RA, Niemegeers JE, Schellekens KHL (1965) Is it possible to product the clinical effects of neuroleptic drugs (major tranquilizer) from animal data. Arzneimittelforsch 15: 104-117

Jellinger K, Riederer P, Gabriel E (1982) D-2-receptor activity in Parkinson's disease and schizophrenia: influence of drug treatment. Arch Pharmacol 321: (Suppl) R 23

Julou L, Bardone MC, Ducrot R, Caffargue B, Loiseau G (1967) Comparison des effects des neuroleptiques dans divers tests, en administration unique et en administrations répétées. Hypothèses sur la signification des tests utilisés et leur valeur prévisionelle. Int Cong Ser 129: 293-303

Jurna J (1968) Depression by antiparkinson drugs of reserpine rigidity. Arch Pharmacol Exp Pathol 260: 80-88

List SJ, Seeman P (1979) Dopamine agonists reverse the elevated ^{3}H-neuroleptic binding in neuroleptic-pretreated rats. Life Sci 24: 1447-1452

Magnus R (1904) Versuche am überlebenden Dünndarm von Säugetieren. Pfluegers Arch Bd 102: 123

Menge HG, Brand U (1978) Zur Differenzierung der Wirkung kataleptogener Stoffe. Arzneimittelforsch 28: 1506-1507

Menge HG, Brand U (1982) Zusammenfassende Darstellung der Pharmakologie von Budipin, einem neuen 4,4-Diphenylpiperidin-Derivat für die Parkinson-Therapie. Arzneimittelforsch 32 (I) 2: 85-98

Menon MK, Vivonica CA, Haddox VG (1984) Evidence for presynaptic antagonism by amantadine of indirectly acting central stimulants. Psychopharmacology 82: 89-92

Nose T, Kogunia M (1970) A simple screening method for antiparkinsonian drugs in mice. Eur J Pharmacol 10: 83-86

Przegalinski E, Bigajska K, Lewandowska A (1982) The effect of budipine on the central serotoninergic system. Pol J Pharmacol Pharm 34: 309-315

Reches A, Burke RE, Kuhn CM, Hassan MN, Jackson VR, Fahn S (1983) Tetrabenazine, an amine-depleting drug, also blocks dopamine receptors in rat brain. J Pharmacol Exp Ther 225: 515-521

Rinne UK, Lonnberg P, Koskinen V (1981) Dopamine receptors in the parkinson brain. J Neural Transm 51: 97-106

Schäfer J, Hackmack G, Eistetter K, Krüger HG, Menge HG, Klosa J (1984) Synthese, physikalisch-chemische Eigenschaften und orientierende pharmakologische Untersuchungen von Budipin und verwandten 4,4-Diphenylpiperidinen. Arzneimittelforsch 34: 233-240

Seeman P, Chan-Wong M, Tedesco J, Wong K (1975) Brain receptors for antipsychotic drugs and dopamine: direct binding assay. Proc Natl Acad Sci USA 72: 4376-4380

Sethy VH, van Woert MH (1973) Antimuscarinic drugs-effect on brain acetylcholine and tremors in rats. Biochem Pharmacol 22: 2655-2591

Sethy VH, van Woert MH (1974) Regulation of striatal acetylcholine concentration by dopamine receptors. Nature 251: 529-530

Stille G (1971) Die Wirkung von Neuroleptica. Arzneimittelforsch 21: 386-527

Sundermann RH, Wooten GE (1980) Biochemical properties of spiperone binding to rat brain. Pharmacology 21: 295-305

The Pharmacodynamics of Budipine on Central Neurotransmitter Systems

J. Offermeier and J.M. van Rooyen

Introduction

In a number of studies, many unpublished and some of them on file in the product information file of the firm Byk Gulden Pharmazeutika Konstanz, it was found that budipine: (a) is an active antiparkinsonian drug, especially against parkinsonian tremor (also against oxotremorine-induced tremors in laboratory animals); (b) enhances the effectiveness of L-dopa treatment in parkinsonism; (c) produces mood elevation; and (d) exhibits a peripheral sympathicotonic action. In studies on the possible mechanism of action various approaches have shown that budipine does not act directly upon a dopaminergic system and that its anticholinergic action is too weak to implicate this mechanism in the antiparkinsonian or antitremor action. The therapeutic results achieved with the drugs thus far are therefore empirical and without a mechanistic explanation. Guided by these empirical findings mentioned and the techniques available to us, we tried to shed some more light on the possible pharmacodynamic mechanisms of budipine. We used two in vivo tests: the locomotor activity (LA) test and the Porsolt swim test, as well as in vitro investigations on synaptosomes prepared from discrete rat brain areas, in our investigation. Both the in vivo tests have clear associations with central catecholaminergic activities and especially the LA model can be manipulated in many ways (by drugs, lesions in certain brain areas, etc.) to give meaningful predictions on pharmacodynamics.

Materials and Methods

Experimental Animals

Male Wistar rats (160-180 g) were used in all experiments.

LA measurements

LA was measured in standard photocell activity cages as described by Offermeier and Potgieter (1975). Three rats were housed per cage and the registrations of activity in at least four cages, i.e. mean activity of twelve rats, were combined for every registogram. All LA experiments were conducted at a temperature of 20° ± 1° C. All drugs tested were dissolved in normal saline and injected intraperitoneally. A 15-min adaptation period was allowed after placing of the rats in the cages before registration of activity was commenced. The results are expressed as mean activity ± standard error of the mean (SEM).

Porsolt Swim Test

The exact procedure described by Porsolt et al. (1977) was followed. Naive rats were individually forced to swim inside vertical acrylic plastic cylinders (height 40 cm, diameter 18 cm) containing 15 cm water of 25°C. After 15 min in the water they were removed and allowed to dry for 15 min in a heated enclosure (32°C) before being returned to their cages. They were replaced in the cylinders 24 h later and the total duration of immobility was measured during a 5-min test period. A rat was judged to be immobile whenever it remained floating passively in the water in a slightly hunched but upright position.

All drugs were administered as a series of three intraperitoneal injections: 24, 5 and 1 h before the 5-min test period on the second day. The results are expressed as the average duration of immobility (mean ± SEM) for groups of six rats for each dose schedule of the test drugs or drug combinations. Differences from control values were assessed for statistical significance using the Dunnett two-tailed test (Dunnett 1955).

Synaptosomal Uptake and Release of Tritiated Neurotransmitters

Crude synaptosomes were prepared from isolated brain areas according to the method of Coyle and Snyder (1969).

Inhibition of Uptake. The synaptosomes were incubated with various concentrations of budipine in Krebs' solution at 37°C for 10 min after which the tritiated neurotransmitters were added and a further 10-min incubation at 37°C allowed. The uptake process was terminated by the addition of Krebs' solution at 0°C. Control experiments included drug-free incubations at 37°C and incubations at 4°C. Following termination of the incubation, the synaptosomes were filtered off onto 0.8 μm Millipore filters at 0°C and washed. The Millipore filters were placed into scintillation vials containing Aquagel and radioactivity was measured in a scintillation counter (Packard Tri-Carb model 3390).

The results are expressed as counts per minute ± standard deviation of the mean. In order to eliminate passive diffusion, the control values at 4°C were subtraced from the values obtained at 37°C in each instance. Thereafter P values (Student's *t*-test) were calculated in order to determine whether the effects obtained in the presence of the various concentrations of budipine differed significantly from the drug-free controls.

Superfusion Release. The synaptosomes were incubated with specific concentrations of the tritiated neurotransmitters at 37°C for 10 min in a Krebs-Ringer bicarbonate (KRB) medium (Mulder et al. 1975). The tritiated neurotransmitter-loaded synaptosomes were then transferred to the chambers of a superfusion system as described by Mulder et al. (1975). The superfusion chambers basically consisted of syringes surrounded by a thermostat bath. The synaptosomal suspension (50 μl) was layered on top of a small column (200 μl) of Sephadex G-15, which was swollen in the KRB medium 18 h previously and superfused with oxygenated KRB medium (either drug-free or containing various concentrations of the test drug) at a rate of 0.25 ml/min, maintained by a peristaltic pump (Gilson Miniplus HP-4); 5-min fractions were collected. The efflux was added to scintillation vials containing Aquagel and the radioactivity was determined as in the experiments of uptake inhibition. The results are expressed as counts per 5 min. For the experiments on KCl-induced release of tritiated neurotransmitters the KRB medium was changed to contain 56 *mM* potassium (KCl replacing an equivalent amount of NaCl).

Results

LA Measurements

The Effects of Budipine on LA. Except for a dose of 10 mg/kg where budipine produced a small increase of LA, budipine at doses of 2.5-25.0 mg/kg did not increase LA in rats (Fig. 1). At higher dosages (50 and 100 mg/kg) budipine appeared to be toxic and several of the rats died.

The Effects of Budipine on the Maximal Increases of LA Produced by Apomorphine and S3608. Apomorphine and S3608 dose-dependently increased LA in rats (results not shown). The maximal increases of LA occurred at 5 mg/kg apomorphine and 30 mg/kg S3608. Pretreatment with budipine (10 mg/kg) attenuated apomorphine (5 mg/kg)-induced LA (Fig. 2). The same budipine pretreatment did not, however, significantly alter S3608 (50 mg/kg)-induced LA (Fig. 3).

The Effects of Clonidine (Alone) on LA. Clonidine at doses up to 2 mg/kg, did not increase LA in rats (Fig. 4).

The Effects of Clonidine on the LA Increases Produced by Apomorphine or S3608. Clonidine at doses of 0.05, 0.10, 1.0 and 2.0 mg/kg potentiated apomorphine (5 mg/kg-induced LA (Figs. 5-8). Clonidine (2 mg/kg) did not significantly alter the LA increase produced by S3608 (30 mg/kg) (Fig. 9).

The Effect of Clonidine on Budipine Attenuation of Apomorphine-Induced LA. Clonidine (0.1 mg/kg) overcame the budipine-attenuation of apomorphine-induced LA (Fig. 10).

The Effect of Budipine on Clonidine Potentiation of Apomorphine-Induced LA. Budipine (10 mg/kg) attenuated the clonidine potentiation of apomorphine-induced LA (Fig. 11). This attenuation was partially overcome by increasing the dose of clonidine to 2 mg/kg (Fig. 12). Complete reversal could not be demonstrated since clonidine at dosages higher than 2 mg/kg appeared to be toxic and the rats died.

Porsolt Swim Test

Budipine dose-dependently decreased the duration of immobility in the Porsolt swim test (Fig. 13).

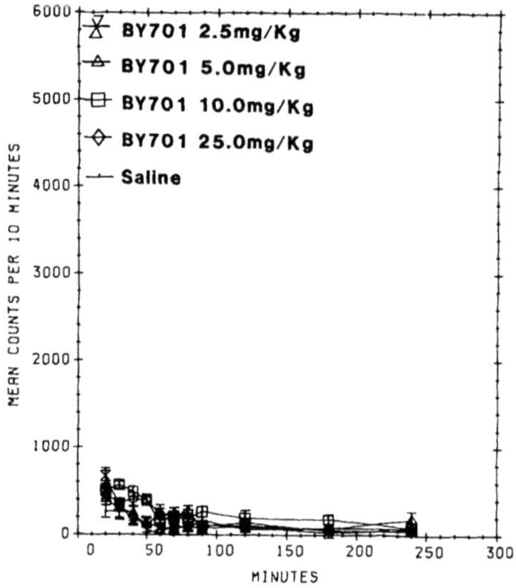

Fig. 1. The effects on increasing intraperitoneal doses of BY 701 (mean ± SEM) on groups of at least 12 rats. No significant increases of LA occurred at dosages up to 25 mg/kg

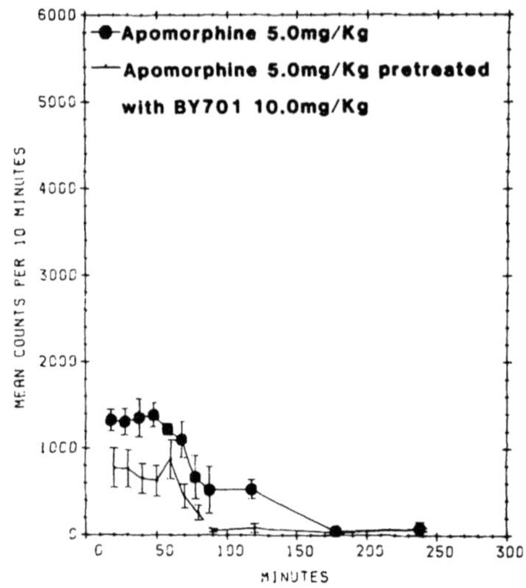

Fig. 2. Antagonism of the effect of apomorphine (5 mg/kg) by the intraperitoneal administration of BY 701 (10 mg/kg) 30 min before apomorphine

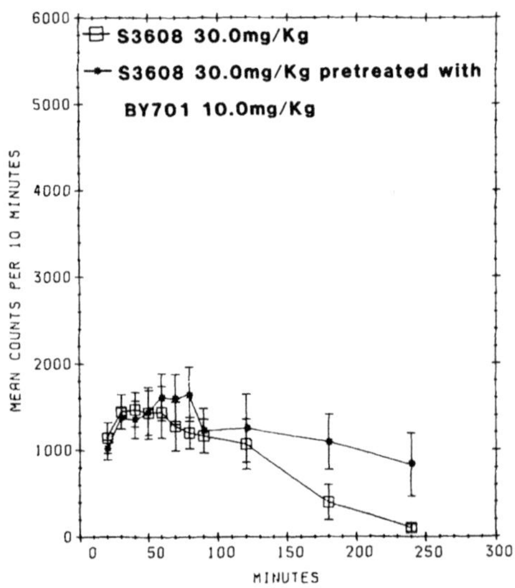

Fig. 3. The effects of BY 701 pretreatment on the LA increases produced by S3608. S3608 (30 mg/kg)-induced LA is not significantly altered by the intraperitoneal administration of BY 701 (10 mg/kg) 30 min before S3608

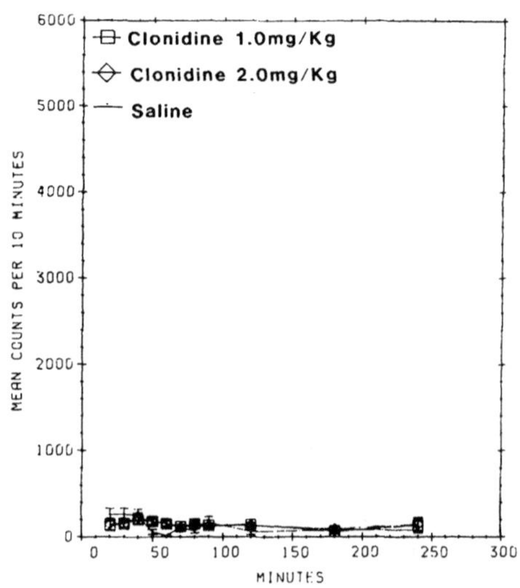

Fig. 4. The effects of increasing intraperitoneal dosages of clonidine on the LA of rats. At dosages of 1.0 and 2.0 mg/kg clonidine alone does not increase LA in rats

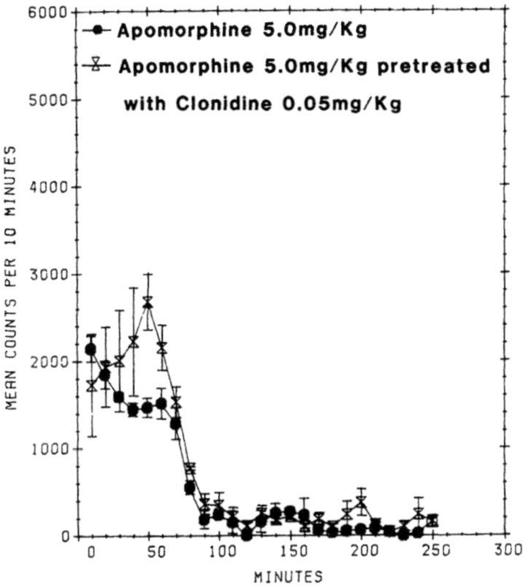

Fig. 5. Potentiation of the effect of apomorphine (5 mg/kg) by the intraperitoneal administration of clonidine (0.05 mg/kg) 30 min before apomorphine

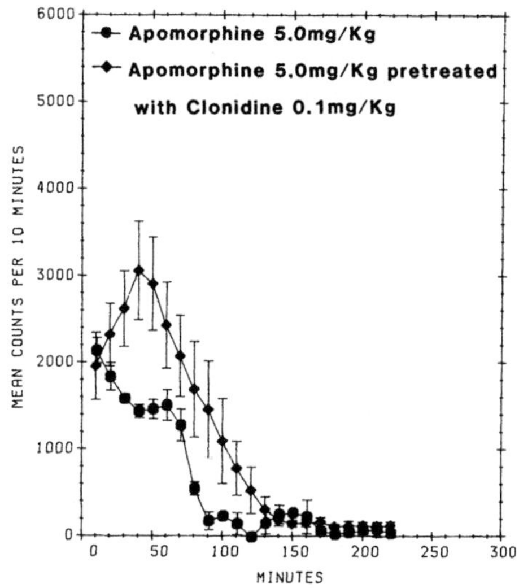

Fig. 6. Potentiation of the effect of apomorphine (5 mg/kg) by the intraperitoneal administration of clonidine (0.1 mg/kg) 30 min before apomorphine

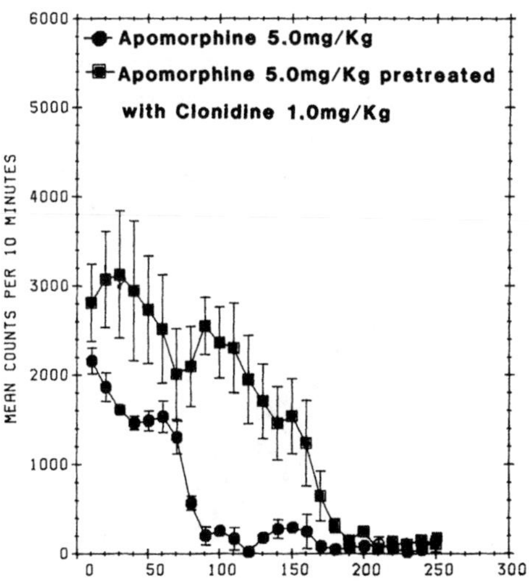

Fig. 7. Potantiation of the effect of apomorphine (5 mg/kg) by the intraperitoneal administration of clonidine (1.0 mg/kg) 30 min before apomorphine

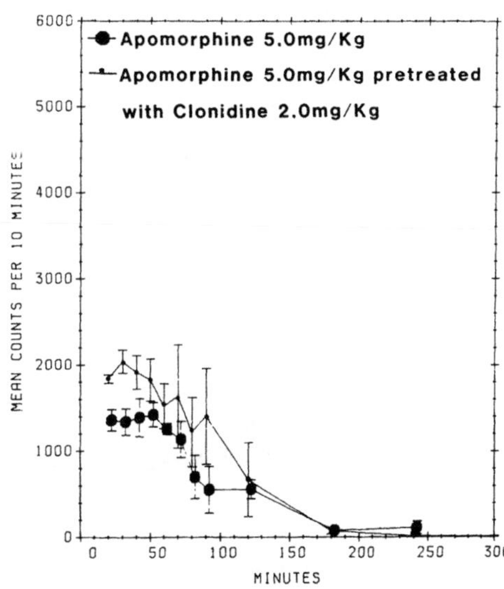

Fig. 8. Potentiation of the effect of apormorphine (5 mg/kg) by the intraperitoneal administration of clonidine (2.0 mg/kg) 30 min before apomorphine

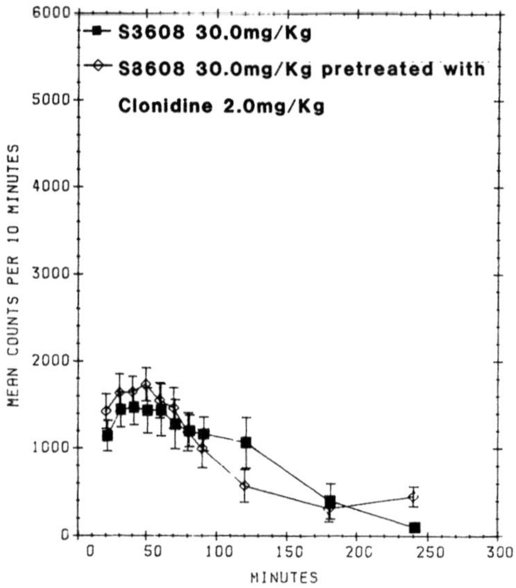

Fig. 9. The effect of clonidine pretreatment on the LA increases produced by S3608 (30 mg/kg)-induced LA is not significantly altered by the intraperitoneal administration of clonidine (2.0 mg/kg) 30 min before S3608

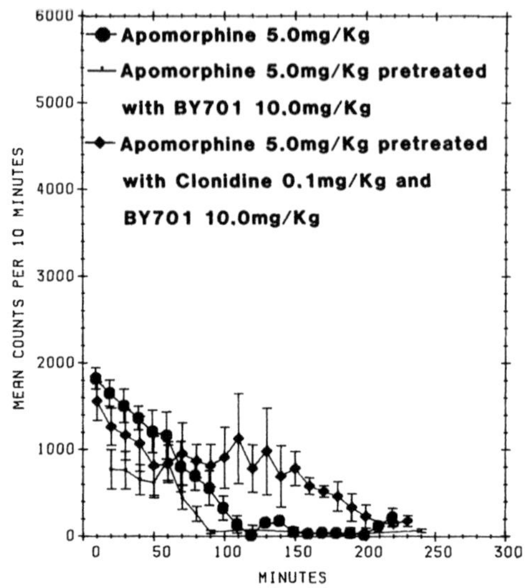

Fig. 10. The effect of clonidine on BY 701 attenuation of apomorphine-induced LA. Intraperitoneal pretreatment with BY 701 (10 mg/kg) 30 min before apomorphine (5 mg/kg) attenuates apomorphine-induced LA. Pretreatment with clonidine (0.1 mg/kg) given simultaneously with BY 701 (10 mg/kg) 30 min before apomorphine, overcomes the BY 701 attenuation of apomorphine-induced LA

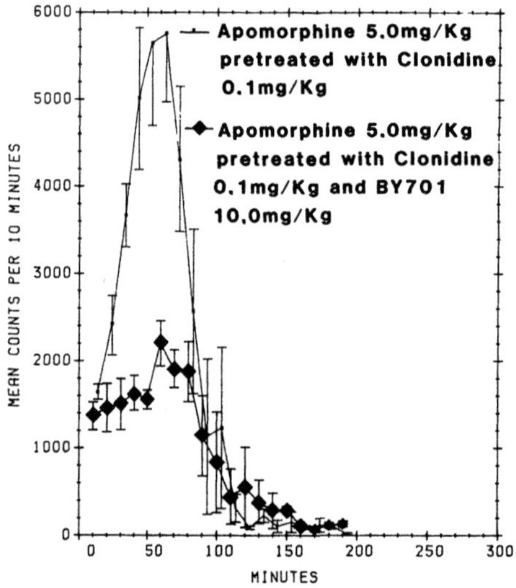

Fig. 11. The effect of BY 701 on the clonidine potentiation of apomorphine-induced LA. Intraperitoneal pretreatment with clonidine (0.1 mg/kg) 30 min before apomorphine (5 mg/kg) potentiates apomorphine-induced LA. Pretreatment with BY 701 (10 mg/kg) given simultaneously with clonidine (0.1 mg/kg) 30 min before apomorphine, attenuates clonidine potentiation of apomorphine-induced LA

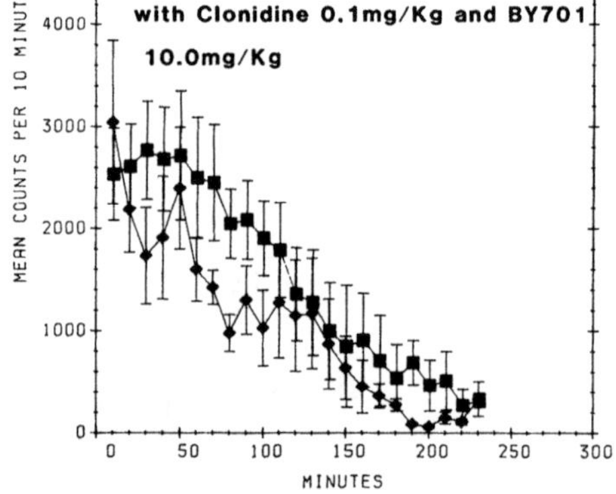

Fig. 12. Partial reversal of the effects of simultaneous intraperitoneal injections of clonidine (0.1 mg/kg) and BY 701 (10 mg/kg) given 30 min before apomorphine (5 mg/kg) by increasing the dose of clonidine to 2 mg/kg

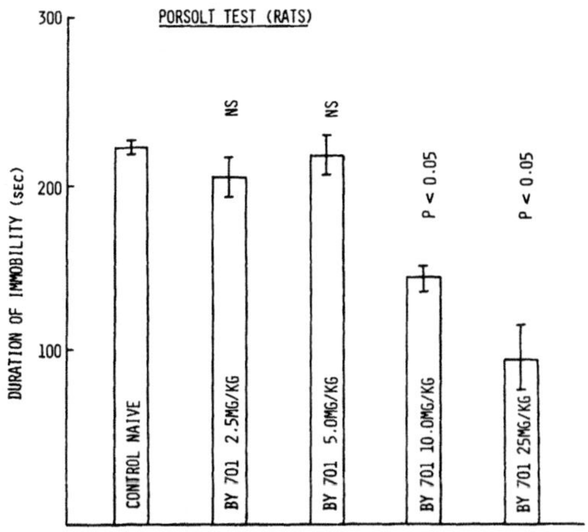

Fig. 13. The effects of increasing dosages of BY 701 on the duration of immobility during the 5-min test period. Each dose indicated for BY 701 was given three times, 24, 5 and 1 h before the 5-min test period. BY 701 dose-dependently decreases the duration of immobility with significant decreases occurring at dosages of 10 and 25 mg/kg

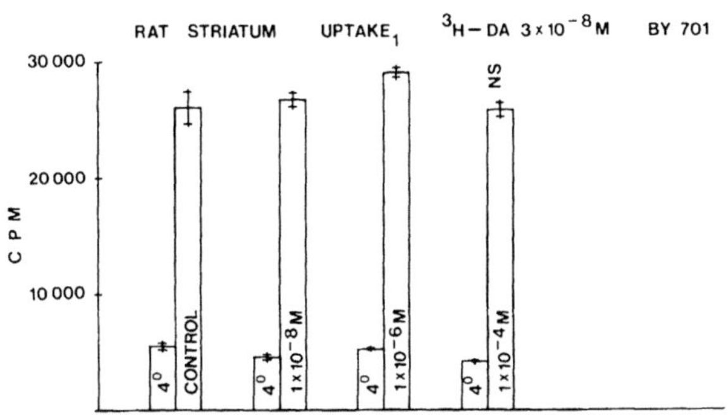

Fig. 14. The effects of BY 701 on the uptake of dopamine ^{3}H in rat striatal synaptosomes. At concentrations up to 10^{-4} *M* BY 701 does not inhibit the uptake of dopamine ^{3}H

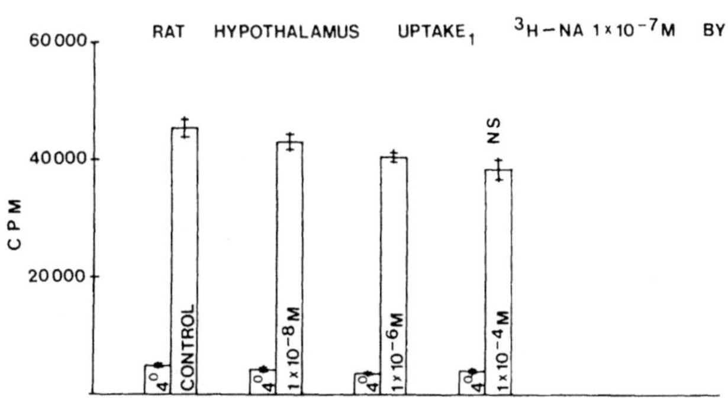

Fig. 15. The effect of BY 701 on the uptake of noradrenaline ^{3}H in rat hypothalamic synaptosomes. At concentrations up to 10^{-4} *M* BY 701 does not inhibit the uptake of noradrenaline ^{3}H

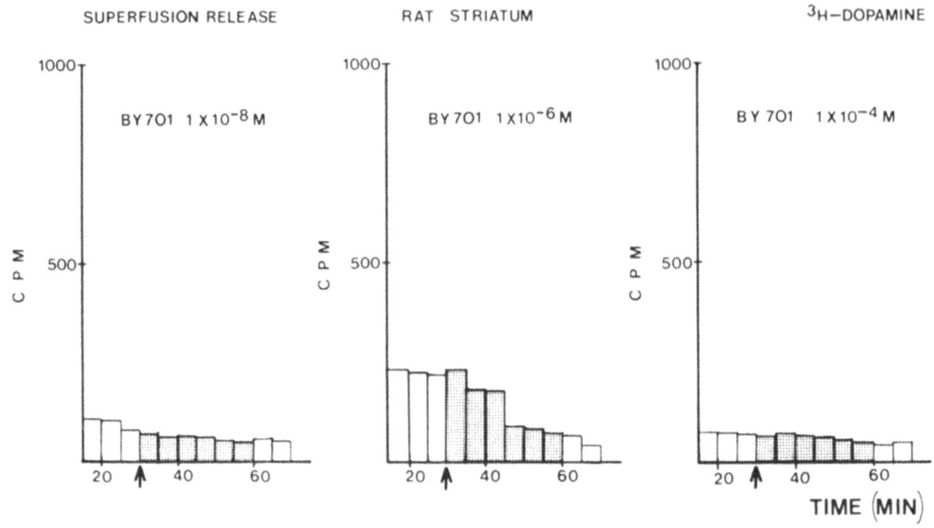

Fig. 16. The effect of BY 701 on the release of dopamine ^3H in rat striatal synaptosomes. BY 701 at concentrations of 10^{-8}, 10^{-6} and 10^{-4} *M* does not cause passive release (i.e., release by displacement) of dopamine ^3H (*arrows* indicate the time of onset of superfusion with BY 701)

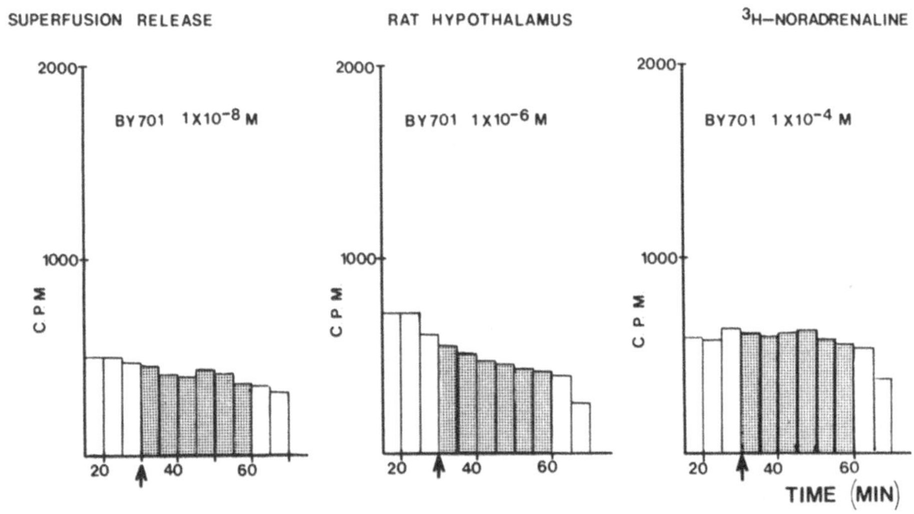

Fig. 17. The effect of BY 701 on the release of noradrenaline ^3H in rat hypothalamic synaptosomes. BY 701 at concentrations of 10^{-8}, 10^{-6} and 10^{-4} *M* does not cause passive release (i.e., release by displacement) of noradrenaline ^3H (*arrows* indicate the time of onset of superfusion with BY 701)

Fig. 19 a, b. The effects of BY 701 on the KCl-induced release of noradrenaline ^3H in rat hypothalamic synaptosomes. KCl (56 *μM*) induces the release of noradrenaline ^3H in rat hypothalamic synaptosomes. At a concentration of 10^{-8} *M*, BY 701 clearly potentiates the KCl-induced release of noradrenaline ^3H (*arrow* in **a** indicates the time of onset of superfusion with 56 *μM* KCl; *arrow* in **b** indicates the time of onset of the simultaneous superfusion with 56 *μM* KCL and 10^{-8} *M* BY 701) ▶

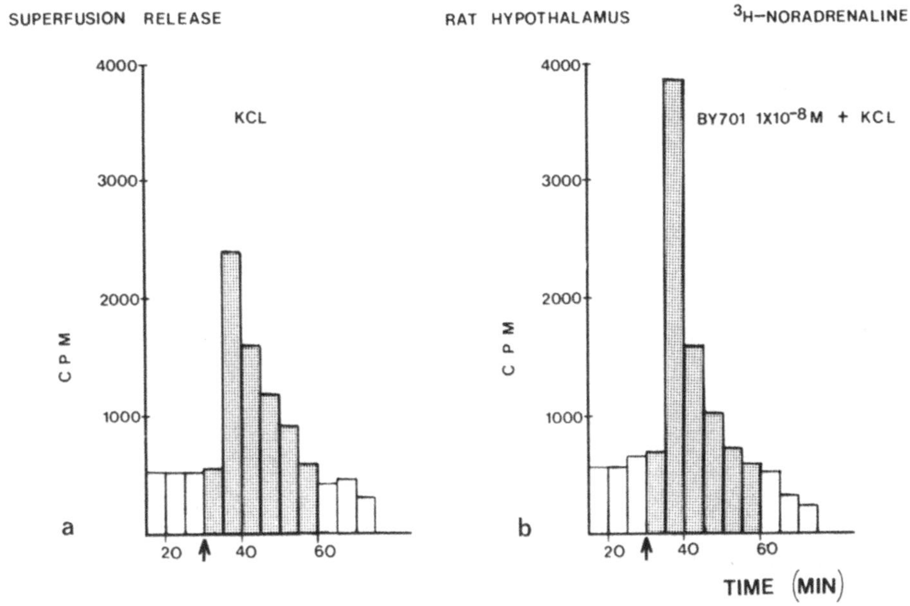

Fig. 18 a, b. The effects of BY 701 on the KCl-induced release of dopamine ^{3}H in rat striatal synaptosomes. KCl (56 *μM*) induces the release of dopamine ^{3}H in rat striatal synaptosomes. At a concentration of 10^{-8} *M*, BY 701 does not influence the KCl-induced release of dopamine ^{3}H (*arrow* in **a** indicates the time of onset of superfusion with 56 *μM* KCl; *arrow* in **b** indicates the time of onset of simultaneous superfusion with 56 *μM* KCl and 10^{-8} *M* BY 701)

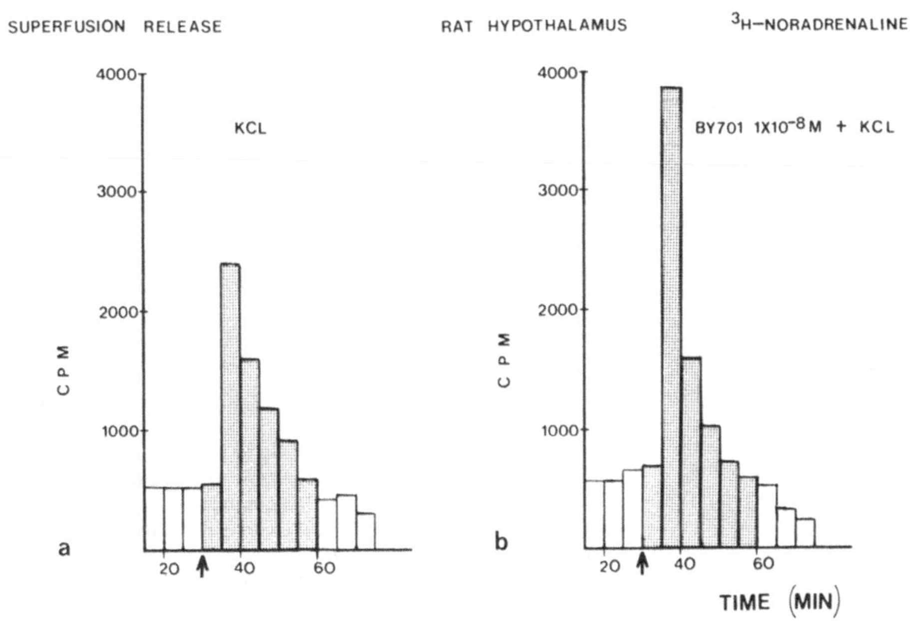
SUPERFUSION RELEASE
RAT HYPOTHALAMUS
^{3}H–NORADRENALINE
KCL
BY701 1X10^{-8}M + KCL
C P M
4000
3000
2000
1000
a
b
20
40
60
TIME (MIN)

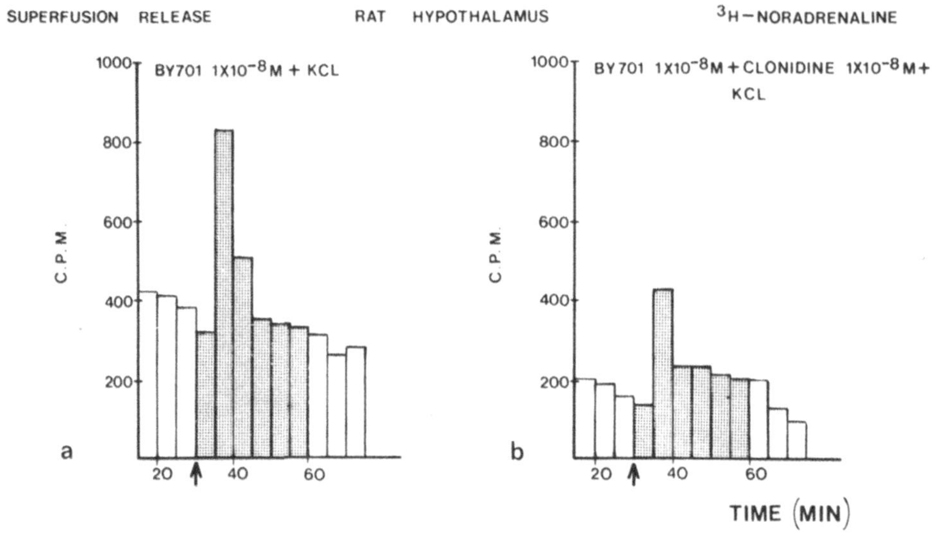

Fig. 20 a, b. The effects of clonidine on the BY 701 potentiation of KCl-induced release of noradrenaline ^{3}H in rat hypothalamic synaptosomes: **a** effect of simultaneous superfusion with KCl (56 μM) and BY 701 (10^{-8} *M*) on the release of noradrenaline ^{3}H in rat hypothalamic synaptosomes; **b** effect of simultaneous superfusion with KCl (56 μM), BY 701 (10^{-8} *M*) and clonidine (10^{-8} *M*) on the release of noradrenaline ^{3}H. By comparing **a** and **b** it is evident that the addition of clonidine attenuates the combined effects of KCl and BY 701 on the release of noradrenaline ^{3}H (*arrows* indicate the time of onset of superfusion)

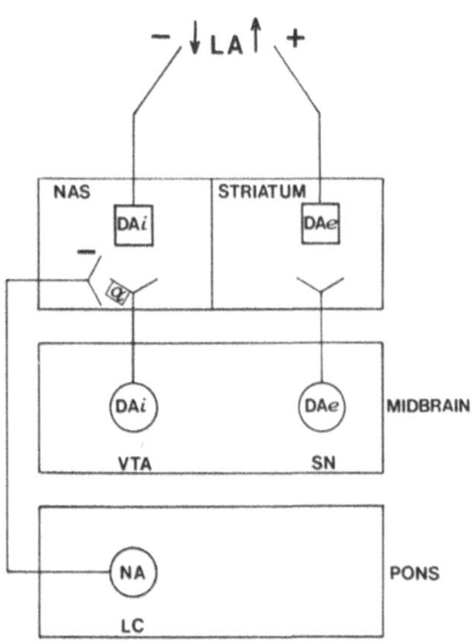

Fig. 21. A diagrammatic presentation of the proposed interactions between dopaminergic and noradrenergic systems in the striatal-accumbens complex (after Cools and van Rossum 1980). For details see text; *DAe* dopamine excitation mediating, *DAi* dopamine inhibition mediation, *LC* locus coeruleus, *NAS* nucleus accumbens speit, *VAT* ventrotegmental area, *SN* substantia nigra

Synaptosomal Studies

The Effects of Budipine on the Synaptosomal Uptake of Tritiated Amines. Budipine at concentrations up to 10^{-4} *M* did not influence the synaptosomal uptake of dopamine ^3H (rat striatal synaptosomes) or noradrenaline ^3H (rat hypothalamic synaptosomes) (Fig. 14 and 15). Budipine, at these same concentrations, also failed to influence the uptake of either GABA ^3H or serotonin ^3H significantly (results not shown).

The Effects of Budipine on the Synaptosomal Release of Tritiated Amines. Budipine at concentrations up to 10^{-4} *M* did not cause passive release (i.e. release by displacement) of dopamine ^3H (rat striatal synaptosomes) or noradrenaline ^3H (rat hypothalamic synaptosomes) (Figs. 16 and 17). Budipine also had no effect on the release of GABA ^3H or serotonin ^3H (results not shown).

The Effects of Budipine on KCl-Induced Release of Tritiated Amines. Budipine did not influence the KCl-induced release of dopamine ^3H in rat striatal synaptosomes (Fig. 18). Budipine (10^{-8} *M*) did, however, potentiate the KCl-induced release of noradrenaline ^3H in rat hypothalamic synaptosomes (Fig. 19). This potentiation was attenuated by clonidine (10^{-8} *M*) (Fig. 20).

Discussion

From the foregoing results it would appear that neither budipine alone nor clonidine (a preferential activator of α_2-adrenoceptors; Langer 1974) alone significantly increased LA in rats. These two drugs did, however, exert opposing effects on apormorphine-induced LA, viz. Budipine attenuated and clonidine potentiated apomorphine-induced LA. The findings that clonidine antagonized budipine attenuation of apomorphine-induced LA and that budipine antagonized the clonidine potentiation of apomorphine-induced LA would furthermore suggest that budipine may act as an antagonist at the receptors where clonidine acts as an agonist, viz. α_2-adrenoceptors. The partial overcoming of the budipine attenuation of clonidine potentiation of apomorphine-induced LA by increasing the dose of clonidine would further support the suggestion of a α_2-antagonistic action of budipine.

It has, however, been reported (Pycock et al. 1977) that clonidine potentiation of apomorphine-induced LA persists after reserpinization in mice. This would suggest that the effects of clonidine on LA are mediated via postsynaptic receptors. Postsynaptic α_2-adrenoceptors have, moreover, been demonstrated in the mammalian brain (U' Prichard et al. 1979).

The finding that budipine potentiated the KCl-induced release of noradrenaline in hypothalamic synaptosomes would suggest that this effect is mediated via blockade of presynaptic receptors, possibly α_2-adrenoceptors. This is substantiated by the demonstration that clonidine attenuated the budipine potentiation of KCl-induced noradrenaline release. The finding that budipine decreased the duration of immobility in the Porsolt swim test may also result from blockade of presynaptic α_2-receptors thereby increasing the concentration of noradrenaline at central synapses. The role of increased concentrations of noradrenaline at central synapses in the decrease of immobility in the swim test is well documented (Porsolt et al. 1979; Borsini et al. 1981).

The exact role of α-adrenoceptors and the interactions between α-adrenoceptors and dopaminergic systems in the mediation of LA in mammals remain to be elucidated. According to Cools and van Rossum (1976, 1980) two dopaminergic systems, a dopamine excitation (DAe) mediating system and a dopamine inhibition (DAi) mediating system, control forward locomotion (LA) in mammals as summarized in Fig 21. The DAe system, subserved by DAe receptors, originates in the substantia nigra and terminates mainly in the striatum. The DAi system, subserved by DAi receptors, originated in the ventrotegmental area and mainly terminates in the accumbens. According to Cools and van Rossum (1976, 1980) a functional balance exists between these two systems and increases of LA are brought about by shift in the balance in favour of DAe dominance. The latter can be brought about by either DAe agonists, viz. apomorphine (Cools and van Rossum 1976), or DAi antagonists, viz. piribedil (Offermeier and van Rooyen 1979) and S3608 (van Rooyen 1980). The terminals of the DAi neurons contain α-like receptors (Fig. 21), activation of which inhibits the release of dopamine onto DAi receptors, thereby shifting the DAe-DAi balance in favour onf DAe dominance. The exact nature of these α-like receptors remain to be elucidated. However, theoretically agonists at these α-like receptors should potentiate DAe agonist-induced LA whereas antagonists should attenuate DAe agonist-induced LA. Similarly, one would expect that agonists or antagonists at these α-like receptors would not markedly alter the maximal increases of LA induced by DAi antagonists. The reason for this is that in the presence of maximal DAi blockade one would expect that manipulation of the α-receptors would only produce a very small further shift in favour of DAe dominance (Fig. 21). It is therefore tempting to speculate that clonidine potentiation and budipine attenuation of apomorphine-induced increases of LA are mediated by agonistic and antogonistic actions, resprectively, at these central α-like receptors.

Conclusions

From the in vivo and in vitro data presented it would appear that budipine acts as an antagonist at central α-like receptors, possibly α_2-adrenoceptors.

Speculations Pertaining to the Effects of α_2-Blockers in Parkinson's Disease

Farley and Hornykiewicz (1976) demonstrated a significant decrease of noradrenaline in several regions, including the locus coeruleus, substantia nigra, nucleus accumbens and posterior hypothalamus of the post-mortem brains of patients with Parkinson's disease. Furthermore, recent experimental evidence suggests that the coeruleus noradrenergic system exerts a synergistic effect on dopaminergically controlled locomotor behaviour; in contrast, the noncoeruleus noradrenergic fibre system (originating from cell groups scattered through the lateral pons-medulla tegmentum) is thought to have an antagonistic effect on the dopaminergic forebrain mechanisms (Hornykiewicz 1980; Donaldson et al. 1978; Kostowski et al. 1978). It would, therefore, be feasible to expect that drugs which increase the release of noradrenaline via an antagonistic action on presynaptic α_2-adrenoceptors may increase both the facilitatory and inhibitory effects of noradrenaline and dopaminergic motor behaviour. However, if the facilitatory noradrenergic system is defective in Parkinson's disease as manifested by the decrease of noradrenaline in the locus coeruleus, one would expect supersensitivity of the α-adrenergic postsynaptic receptors subserving this system, and therefore dominance, after administration of drugs which increase the release of noradrenaline.

It has furthermore been demonstrated that clonidine (0.1 mg/kg given intraperitoneally) reduces the striatal turnover of dopamine whereas the α_2-blocker yohimbine increases striatal dopamine turnover (Anden and Grabowska 1976). This provides another way in which α_2-antagonists may alleviate the symptoms of Parkinson's disease. The finding, however, that α_2-antagonists decrease apomorphine-induced LA in vivo in rats, is difficult to reconcile with an antiparkinsonian effect of this type of compound. The latter would suggest that degeneration of the nigrostriatal pathways and/or the locus coeruleus accumbens pathways, possibly owing to altered sensitivities of α-adrenoceptors and dopaminergic receptors, may modify the action of α_2-agonists or α_2-antagonists.

References

Anden N-E, Grabowska M (1979) Pharmacological evidence for a stimulation of dopamine neurons by noradrenaline neurons in the brain. Eur J Pharmacol 39: 275-282

Borsini F, Bendotti C, Velkov V, Rech R, Samanin R (1981) Immobility test: effects of 5-hydroxytryptaminergic drugs and role of catecholamines in the activity of some antidepressants. J Pharm Pharmacol 33: 33-37

Cools AR, van Rossum JM (1976) Excitation-mediating and Inhibition-mediating dopamine receptors: A new concept towards a better understanding of electrophysiological, biochemical, pharmacological, functional and clinical data. Psychopharmacologia (Berlin) 45: 243-244

Cools AR, van Rossum JM (1980) Multiple receptors for brain dopamine in behavioural regulation: Concept of dopamine-E and dopamine-I receptors. Life Sci 27: 1237-1253

Coyle JT, Snyder SH (1969) Catecholamine uptake by synaptosomes in homogenates of rat brain: stereospecificity in different areas. J Pharmacol Exp Ther 170: 221-231

Donaldson I macG, Dolphin AC, Tenner P, Pycock C, Marsden CD (1978) Rotational behaviour produced in rats by unilateral electrolytic lesions of the ascending noradrenergic bundles. Brain Res 138: 487-509

Dunnett CW (1955) New tables for multiple comparisons with a control. J Am Stat Assoc 50: 1986-1121

Farley IJ, Hornykiewicz O (1976) Noradrenaline in subcortical brain regions of patients with Parkinson's disease and control subjects. Roche, Basel, pp 178-185

Hornykiewicz O (1980) Biochemical abnormalities in some extrastriatal neuronal systems in Parkinson's disease. In: Rinne UK, Klingler M, Stamm G (eds) Parkinson's disease: Current progress, problems and management. Elsevier/North-Holland, Amsterdam, pp 109-119

Kostowski W, Jerlicz M, Bidzinski A, Hauptman M (1978) Evidence for existence of two opposite noradrenergic brain systems controlling behaviour. Psychopharmacologia (Berlin) 59: 311-312

Langer SZ (1974) Presynaptic regulation of catecholamine release. Biochem Pharmacol 23: 1793-1800

Mulder AH, Van den Burg WB, Stoof JC (1975) Calcium-dependent release of radiolabelled catecholamines and serotonin from rat brain synaptosomes in a superfusion system. Brain Res 99: 419-424

Offermeier J, Potgiererer B (1975) Some effects of fenfluramine and its derivatives on central catecholaminergic systems of mice. Postgrad Med J 51: 65-71

Offermeier J, Van Rooyen JM (1979) Pharmacological evidence for the activity of piribedil on mesolimbic dopaminergic pathways. Psychol Med (B) 11: 187-204

Porsolt RD, Le Pichon M, Talfre M (1977) Depression: a new model sensitive to antidepressant treatments. Nature 255: 730-732

Porsolt RD, Bentin A, Blavek N, Deniel M, Talfre T (1979) Immobility induced by forced swimming in rats: effects of agents which modify central catecholamine and serotonin activity. Eur J Pharmacol 57: 201-210

Pycock CJ, Tenner PG, Marsden CD (1977) The interaction of clonidine with dopamine-dependent behaviour in rodents. Naunyn Schmiedebergs Arch Pharmacol 297:133-141

U'Prichard DC, Bechtel WD, Rouot BM, Synder SH (1979) Multiple apparent α-noradrenergic receptor binding sites in rat brain: Effect of 6-hydroxydopamine. Mol Pharmacol 16: 47-60

Van Rooyen JM (1980) Drugs and the mesotelencephalic system. Doctoral thesis, Potchefstroom University

Biochemical and Pharmacologic Aspects of the Mechanism of Action of Budipine*

H. Przuntek and J.-P. Stasch

Budipine is of clinical value in the treatment of severe cases of migraine, in a particular group of patients with essential tremor and as a therapeutic adjuvant in Parkinsons's disease. In reserpine-induced Parkinson syndrome, budipine brings about an increase in the levels of biogenic amines in the nucleus caudatus and also in the whole brain. Evidence that this increase was brought about monamine oxidase (MAO) or catechol O-methyltransferase (COMT) inhibition could not be found. Rather it is suggested that the increase is due to a reuptake inhibition.

The low receptor binding affinities of budipine rule out the possibility that it is a dopaminergic, serotonergic, GABAergic or benzodiazepinergic agonist. Whether the medium affinity to the opiate receptor is of clinical significance is still not clear.

Introduction

L-Dopa, certainly still the most effective antiparkinsonian drug, in long-term therapy produces side effects such as dyskinesias and on-off phenomena. For this reason, we saw at an early stage the need for clinical tests of other antiparkinsonian substances. Since 1974 we have been investigating a number of diphenylpiperidines including budipine (Gerlach et al. 1982, 1983; Neuser et al. 1983; Przuntek and Stumptner 1975; Przuntek et al. 1977; Seemann et al. 1976).

Budipine is a butyldiphenylpiperidine which, according to the physiologic-pharmacologic studies of Menge and Brand (1982), is an antiparkinsonian substance with a new action profile.

Clinically, we have observed beneficial effects of budipine in the prophylactic therapy of migraine (Przuntek, to be published a, b) and of essential tremor and as a therapeutic adjuvant in the treatment of Parkinson's disease. In addition, we feel that the long-term side effects of L-dopa, such as the on-off phenomenon and hyperkinesias, could be reduced along with the dosage of L-dopa or dopaminergic agents when used in combination with budipine. Budipine has less tendency to induce hyperactivity than its structural analogue prodipine and so with budipine anxious agitation states are a less frequent side effect.

In animal experiments, budipine has an anti-haloperidol, anti-oxotremorine, anti-reserpine, and anti-tetrabenazine effect (Menge and Brand 1982) which suggests that it could be an effective antiparkinsonian substance. Experiments with turning rats and

* Supported by the Deutsche Forschungsgemeinschaft (P 145/3/1).

doses up to 10 mg/kg have shown that budipine is not a classic dopaminergic agent, unlike apomorphine or bromocriptine, although it enhances the effect of L-dopa. It is striking that budipine has no pharmacologic effects worth mentioning on normal animals, only on those with pharmacologically provoked model disorders.

In spite of very detailed physiologic-pharmacologic studies, Menge and Brand (1982) did not provide an adequate explanation for the effect of budipine on Parkinson's disease. We have attempted to clarify the situation with biochemical-pharmacologic methods. To this end we have measured the binding affinities of budipine to various receptor types in the brain. In addition, we have investigated the effect of budipine on the brain levels of biogenic amines and their metabolites.

Methods

Receptor Preparation

The preparation of receptors and the measurement of the binding affinities of budipine to the opiate, dopaminergic, serotonergic, benzodiazepinergic, and GABAergic receptors were carried out as described in the literature (Baraczka and Sperk 1981; Chang et al. 1979; Kuhn et al. 1981; Neuser et al. 1983; Petroutka and Snyder 1979; Seemann et al. 1976).

Estimation of Biogenic Amines in Rat Brain

After homogenization of the rat brain (male Sprague-Dawley rats, 250-300 g) and fractination of the homogenate on strongly acid cation exchangers, the levels of biogenic amines were measured fluorimetrically by the method of Atack and Magnusson (1978).

High Pressure Liquid Chromatography

High pressure liquid chromatographic (HPLC) separation and quantitative estimation with an electrochemical detector of dopamine, serotonin, 5-hydroxyindoleacetic acid (HIAA), homovanillic acid (HA), and 3,4-dihydroxyphenylacetic acid (DOPAC) in the nucleus caudatus was carried out after homogenization of the tissue in the elution buffer using a Milton Roy HPLC apparatus (Constametric pump Model III, chromatography control module CCM, calculator-integrator No. 301, printer-plotter, rheodyne injection system with a 200-μl injection loop). Chromatography column 250x4.6 mm i.d. filled with Shandon Hypersil ODS 5 μm; precolumn 20x4.6 mm i.d. filled with LiCrosorb RP 10 μm (Bischoff Analysentechnik, Stuttgart, Federal Republic of Germany). Electrochemical detector Methrom, model 656/64 1. The potential was set at 0.75 V relative to a reference electrode with a sensitivity of 50-100 nA/V. Composition of the mobile phase 0.1 *M* sodium citrate, 75 m*M* disodium phosphate, 0.75 m*M* sodium heptansulfonate, and 14% methanol, pH 3.9. The sample was applied directly in the elution buffer. 5-hydroxyindole-2-carbonic acid was used as internal standard (Gerlach et al. 1983).

Results

In the receptor binding studies, is was shown that the budipine receptor binding affinity, expressed as the inhibitor concentration for half-maximum ligand binding, IC_{50}, to the neuroleptic receptor (spiroperidol ^3H) was 0.12 m*M*, to the dopaminergic receptor (apomorphine ^3H) 0.55 m*M*, to the serotonergic receptor (serotonin ^3H) 0.76 m*M*, to the GABAergic receptor complex (muscimol ^3H) 1.9 m*M* and (diazepam ^3H) 31 m*M* and to the opiate receptors (D-Ala^2-D-Leu^5-enkephalin ^3H) 0.026 m*M*, (dihydromorphine ^3H) 0.020 m*M*, and (etorphine ^3H) 0.059 m*M*. Thus, the binding affinity of budipine to the serotonergic, dopaminergic, GABAergic and benzodiazepinergic receptors is low, to the opiate receptors, medium.

Since budipine brought about a marked improvement in the mobility of reserpinized rats, we investigated the effect of budipine on biogenic amine levels after reserpine treatment. Pretreatment of the experimental animals with budipine, 10 mg/kg i.p., induced no change in the levels of the biogenic amines noradrenaline, dopamine, serotonin, and histamine in whole rat brain within 50 h of its application (Fig. 1).

Treatment of the rats with reserpine, 0.3 mg/kg i.p., 50 h before decapitation, led to a significant drop in the brain levels of biogenic amines. When this dose of reserpine was supplemented with budipine, 10 mg/kg i.p., a marked rise in biogenic amine levels resulted, the maximum being reached 80-320 min after budipine injection. Similarly, in the nucleus caudatus after treatment of rats with budipine, 15 mg/kg i.p., no change in serotonin or dopamine levels was found. The concentrations of HIAA, HVA, and DOPAC in the nucleus caudatus were also unchanged so that MAO inhibition or COMT inhibition are unlikely. After treatment with reserpine, 1.5 mg/kg i.p., 16 h before decapitation, a significant decrease in serotonin and dopamine levels and an increase in the levels of their metabolites could be demonstrated. A supplementary injection of budipine, 15 mg/kg i.p., the amount which must be considered as the maximum therapeutic dose in animal experiments, leads to a marked increase in the levels of these biogenic amines without any significant change in the levels of their metabolites (Table 1).

Discussion

Menge and Brand (1982) concluded from their animal studies that budiphine had pronounced effect on the motor behavior of pharmacologically modified animals. Our biochemical studies demonstrate that budipine treatment does not lead to an alteration in the rat brain levels of the biogenic amines serotonin, dopamine, noradrenaline, and histamine.

The unchanged levels of serotonin and dopamine and of their metabolites HIAA, HVA, and DOPAC in the nucleus caudatus after budipine treatment suggest that budipine does not inhibit MAO or catecholo-methyltransferase to any significant extent. Reserpine treatment of experimental animals induces motor behavior which mimics Parkinson's disease. Reserpinization of experimental animals induces a significant reduction in the brain levels of the biogenic amines dopamine, noradrenaline, serotonin, and histamine. Subsequent treatment with budipine leads to an increase in amine levels.

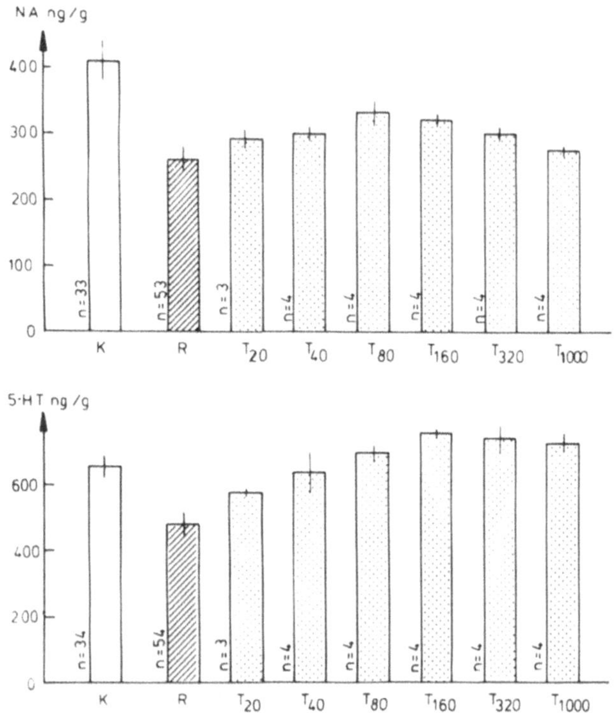

Fig. 1. Time course of the effect of budipine and reserpine on the noradrenaline (NA), dopamine (DA), serotonin (5-HT), and histamine (HM) levels (ng per gram fresh weight) in the rat brain. K controls (0.9% NaCl solution i.p.); R 1x0.3 mg/kg reserpine i.p. 50 h before decapitation; T_x 1x0.3 mg/kg reserpine i.p. 50 h before decapitation and 1x10.0 mg/kg budipine i. p. x min before decapitation

Table 1. The effect of reserpine and budipine on the concentrations of DA, DOPAC, HVA, 5-HT, and HIAA in the nucleus caudatus of the rat brain[a]

Substance	Number of rats	Concentration (ng per milligram protein)				
		DA	DOPAC	HVA	5-HT	HIAA
Physiological saline	12	88.5±7.0	12.0±0.8	6.7±0.7	6.9±0.6	7.3±0.5
Budipine	4	89.1±6.3	12.2±0.8	6.8±1.2	6.8±0.7	7.1±1.1
Reserpine	5	18.8±3.8	14.1±1.7	10.5±0.3	2.9±0.6	8.7±0.9
Reserpine and budipine	5	50.3±9.5	14.0±2.5	10.0±0.6	4.8±0.7	10.1±0.7

[a] Quantitative estimation was carried out by high pressure liquid chromatographic separation and electrochemical detection (Gerlach et al. 1983). Rats were decapitated 16 h after reserpine injection (1.5 mg/kg i.p.) and 100 min after budipine injection (15 mg/kg i.p.). Values are means ± standard deviation in ng/mg protein.

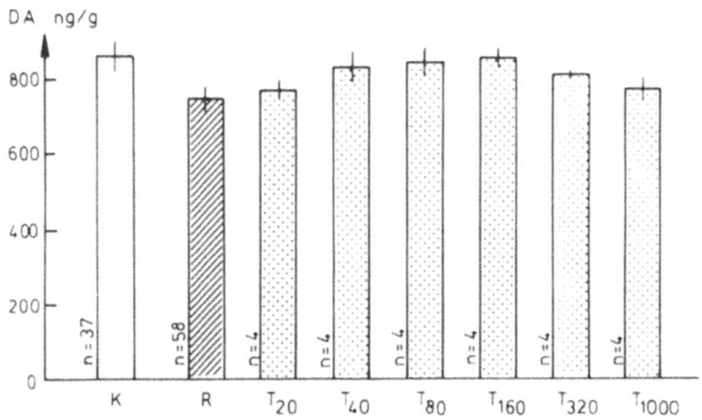

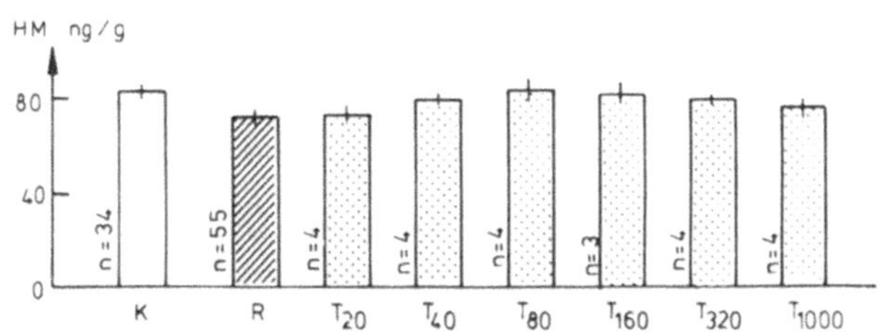

These results suggest that budipine, like the diphenylpiperidine, aprotifin, inhibits the reuptake of the biogenic amines into the presynaptic neuron. This is in agreement with the findings of Eltze (1980) who was able to demonstrate an inhibition by budipine of serotonin reuptake in thrombocytes, whoch provides a possible explanation for its effectiveness in migraine therapy (Przuntek, to be published a, b) The binding affinity of budipine to the dopaminergic receptors D2 and D3 as well as to the serotonergic, GABAergic, and benzodiazepinergic receptors is so low that a direct stimulation of these receptors can be ruled out. To what extent stimulation of the opiate receptors plays a clinical role requires further study.

Acknowledgments. We should like to thank Mr. Jürss for linguistic help and Ms. Srutek and Ms. H. Eckardt for technical assistance. Further we thank Byk Gulden Lomberg GmbH, Federal Republic of Germany, for generous supplies of budipine and Sprague-Dawley rats.

References

Atack C, Magnusson T (1978) A procedure for the isolation of noradrenaline (together with adrenaline), dopamine, 5-hydroxytryptamine and histamine from the same tissue sample using a single column of strongly acidic cation exchange resin. Acta Pharmacol Toxicol (Copenh) 42: 35-57

Baraczka K, Sperk G (1981) Measurement of gammaaminobutyric acid in human cerebrospinal fluid. Radioreceptor assay using ^{3}H muscimol. Clin Chim Acta 109: 77-82

Chang KJ, Cooper BR, Hazum E, Cuatrecasas P (1979) Multiple opiate receptors: Different regional distribution in the brain and differential binding of opiates and opioid peptides. Mol Pharmacol 16: 19-104

Eltze M (1980) The effects of prodipine and budipine on ^{14}C-5-hydroxytryptamine uptake and release by human blood platelets. Arzneimittelforsch 30: 1129-1132

Gerlach M, Jutzi P, Stasch J-P, Przuntek H (1982) Synthese und pharmakologische Eigenschaften von 4,4-Diphenly-4-sila-piperidinen. Z Naturforsch 376: 657-662

Gerlach M, Jutzi P, Stasch J-P, Przuntek H (1983) Synthese und pharmakologische Eigenschaften von silylierten Dopaminen und 4,4-Diphenylpiperidinen. Z Naturforssch 38b: 237-242

Johnson RP, Sar M, Stumpf W (1980) A topographic localization of enkephalin on the dopamine neurons of the rat substantia nigra and ventral tegmental area demonstrated by combined histofluorescenceimmunocyto chemistry. Brain Res 194: 566

Kilts CD, Breese GR, Maitman RB (1981) Simultaneous quantification of dopamine, 5-hydroxytryptamin and four metabolically related compounds by means of reversed-phase high-performance liquid chromatography with electrochemical detection. J Chromatogr 225: 347

Kuhn W, Neuser D, Przuntek H (1981) ^{3}H-Diazepam displacing activity in human cerebrospinal fluid. J Neurochem 37: 1045-1047

Menge HG, Brand U (1982) Zusammenfassende Darstellung der Pharmakologie von Budipin einem neuen 4,4-Diphenyl-piperidinderivat für die Parkinson-Therapie. Arzneimittelforsch 32: 85-98

Neuser D, Stasch JP, Witteler M, Kuhn W, Gerlach M, Jutzi P, Przuntek H (1983) The interaction of l-alkyl-4,4-diphenylpiperidines with opiate receptors. Eur J Pharmacol 87: 315-318

Petroutka SJ, Snyder HS (1979) Multiple serotonin receptors: differential binding of (^{3}H) 5-hydroxytryptamine, (^{3}H)-lysergic acid diehtylamide and (^{3}H)-spiroperidol. Mol Pharmacol 16: 687-699

Przuntek H (to be published a) Importance of Budipine in the treatment of migraine. J. Neurol

Przuntek H (to be published b) Vergleichende Untersuchung zur Behandlung der Migräne mit der Substanz FU-127 und By 701. Fortschr Neurol Psychiatr

Przuntek H, Stumptner K (1975) Influence of Di-N-propylacetate and prodipine on extrapyramidal disorders and the DA,NA,5-HT,HM and GABA content of rat brain. In: Proceedings of the 6th International Congress of Pharmacology, p 543

Przuntek H, Pflughaupt KW, Stumptner K, Kestler P (1977) Influence of Isopropyl-Diphenyl-Piperidine on Parkinson's disease and DA,NA,5HT,GABA,ACH and HM content of rat brain. Excerpta Med 427: 204

Seemann P, Lee T, Chau-Wong M, Wong K (1976) Antipsychotic drugs doses and neuroleptic/dopamine receptors. Nature 261: 717-718

Stasch JP (1983) Silapiperidine und silylierte Piperidine mit potentieller Antiparkinsonaktivität-Synthesen, Rezeptorbindungsstudien und Einflüsse auf die chemische Signalübertragung im Zentralen Nervensystem. Medical Dissertation, University of Würzburg

Pharmacokinetics and Metabolism of Budipine in Animals and Humans

K. Zech, E. Sturm and G. Ludwig

Introduction

Budipine (Hackmack et al. 1979) has been proven as an effective drug in experiments in animals (Menge and Brand 1982) as well as in clinical investigations for treating patients (Siegfried 1982) suffering from parkinsonian syndrome. The present report includes a summary of the studies hitherto conducted on the pharmacokinetics and metabolism of budipine.

Materials and Methods

Synthesis of Budipine ^{14}C

For studies in the rat and dog, labeled ^{14}C budipine was synthesized. The investigations were carried out with budipine labeled in positions 3 and 5 of the piperidine residue, synthesized in our laboratories.

Analytic Procedures for Quantitative Determination of Budipine and Metabolites in Biologic Sample Material

Radioactivity Measurements. Quantitative measurement of radioactivity of samples was done in scintillation spectrometers with external standardization for quench correction. In studies for measurements of organ distribution of radioactivity, one aliquot of the homogenized sample was burned and the $^{14}CO_2$ engendered was collected in the scintillator mixture to determine radioactivity.

Determination of Budipine and Metabolites in Urine by HPLC. Figure 1 shows a typical high pressure liquid chromatogram (HPLC), as it was obtained after injection of a rat urine sample after extraction with dichloromethane/isopropanol (9/1 v/v). The separation procedure used allows for simultaneous determination of unchanged budipine and metabolite M1.

The conjugate of M1 was identified by incubation of the aqueous layer (after extraction) with either arylsulfatase or β-glucuronidase or glusulase followed again by extraction for HPLC separation. For identification of metabolite M1, human and rat urine extracts after administration of radiolebeled budipine were analyzed with continuous

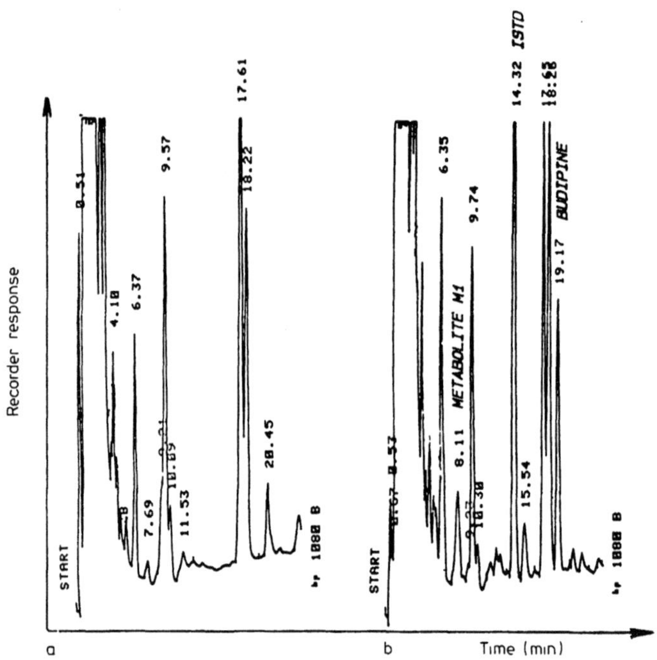

Fig. 1 a, b. Typical high pressure liquid chromatogram of a sample from the 8 to 24-h rat urine **a** blank value; **b** chromatogram after 10 mg/kg. The peaks in the chromatogram correspond to the following concentrations: budipine 1.394 mg/l (279 ng); metabolite M1 1.522 mg/l (304 ng); internal standard 2.52 mg/l (504 ng). Chromatographic conditions: mobile phase: 20 m*M* ($NH_4H_2PO_4$, pH 2.7 with H_3PO_4, for 5 min under isocratic conditions with 30% acetonitrile, then gradient elution with increasing content of acetonitrile (0.5%/min); flow 0.8 ml/min. Oven temperature 35° C. Stationary phase: Nucleosil 5 C-18, 100x3.2 mm. Detection at 220 nm

UV and radioactivity detection of the column effluent. Cochromatography of the urine extract with synthetic metabolite was done additionally.

Quantitave Gas Chromatographic Mass Spectrometric Determination of Budipine and Metabolite M1 in Human Serum. The analytic procedure (Fig. 2) was optimized for determining budipine and its metabolite in human serum at the nanogram level. The method involves addition of internal standards (isopropyl analogues of budipine and metabolite) to 1 ml serum, followed by extraction with dichloromethane/isopropanol (9/1 v/v), derivatization of metabolite M1 with bistrimethylsilyltrifluoracetamide (BSTFA), and gas chromatographic separation. Specifity is provided by selected ion recording of the characteristic $M\text{-}CH_3^+$ ions formed upon electron impact ionization. Peak area ratios, used to calculate serum concentrations, were obtained from digitally recorded mass chromatograms.

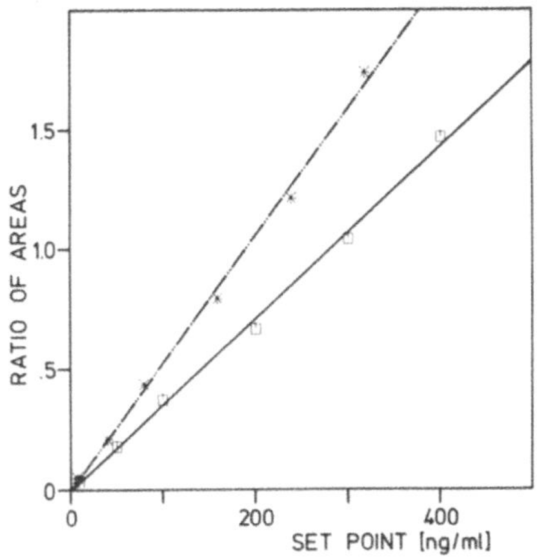

Fig. 2. Standard calibration curve for the quantitative detection of budipine and metabolite M1 in serum by gas chromatography-mass spectroscopy: Regression equation for budipine $y = 0.00360x - 0.011$ ($r = 0.9984$); for metabolite M1 $y = 0.00533x - 0.0099$. Ratio of areas: area budipine (m/z 278)/area internal standard (m/z 264) or area metabolite (m/z 366)/area internal standard (m/z 352)

Results

Pharmacokinetics in the Rat

Concentrations in the Blood. After intravenous administration, a biphasic course of the blood concentration is measured, with the initial distribution phase almost terminated at about 0.5 h. Thus, for example, the blood level 1 h following bolus injection falls to less than 10%, after 4-5 h to about 1% of the concentration measured 5 min after administration. This extensive decline in blood concentration suggests that distribution is rapid and tissue absorption is good. The course of activity indicates further that a metabolite with a half-life longer than that of budipine is present, a result which was confirmed by thin layer chromatographic separation of radioactive components: Thus, 1 h after bolus injection, total radioactivity consisted of about 40% of parent compound and 40% of conjugated metabolite M1 (as major constituents), while 10 h after administration the constitution is 8% and 66% (budipine and conjugated metabolite, respectively).

After oral administration to rats, blood level peaks between 4 and 8 h followed by elimination of total activity with a half-life of about 7 h. This applies to both routes of administration (Fig. 3), but this half-life characterizes the elimination of metabolites rather than that of parent compound. Chromatographic separation of radioactive compounds confirms this; 1 h after oral administration only 24% of radioactivity belongs to budipine while 65% is due to conjugated metabolite M1, but 10 h after administration the ratio is 7.6% to 80% (budipine and conjuagted metabolite, respectively).

Calculation of the extent of absorption on the basis of blood data is difficult with compounds having high affinity to organs or tissue because blood concentrations reflect only a few percent of the administered dose. This may explain the fact that on the basis of urinary and biliary data, absorption is nearly 100%, while on the basis of comparing the areas under the concentration-time curves after both forms of administration according to Dost (1968) absorption is only 61%.

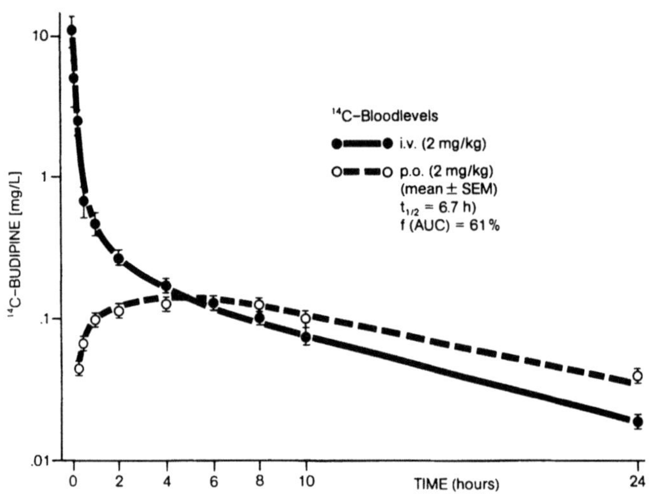

Fig. 3. ^{14}C blood level after single administration of budipine ^{14}C to rats (2 mg/kg i.v. and p.o., N = 10)

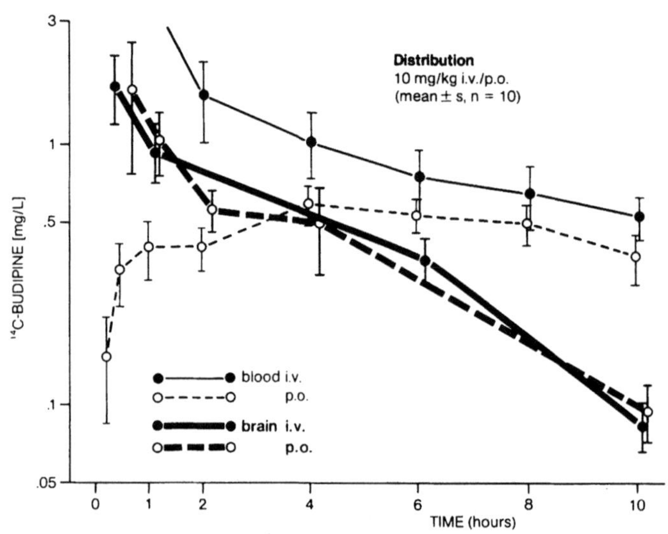

Fig. 4. Comparison of mean blood and brain level of total activity, following administration of 10 mg/kg budipine ^{14}C to rats. N = 10 for blood level and brain p.o. and N = 5 for brain i.v.

Distribution of Budipine. The rapid distribution of budipine ^{14}C was confirmed by means of intact animal autoradiography in the rat. Budipine shows high affinity to liver, lung, kidney, spleen as well as to stomach and intestine wall and secretory glands.

The penetration of the blood-brain barrier was studied by comparing radioactivity measurements in the blood and brain homogenates. Independant of the route of ad-

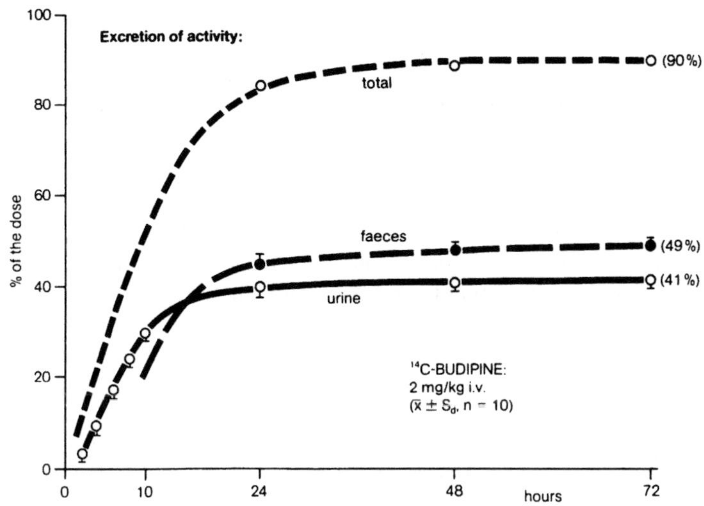

Fig. 5. Excretion of total radioactivity, and radioactivity in urine and feces of rats after a single intravenous dose of 2 mg/kg budipine ^{14}C (mean ± standard deviation, $N = 10$)

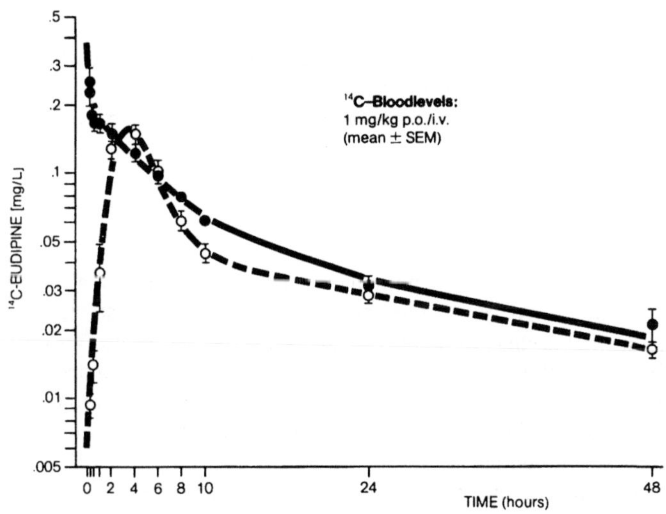

Fig. 6. ^{14}C blood level after single oral and intravenous administration of 1 mg/kg budipine ^{14}C to dogs (mean ± standard error, $N = 4$)

ministration, very rapid and high activity concentrations can be determined in the brain of the rat (Fig. 4). Surprisingly, activity concentrations following oral administration are higher in the brain than in the blood up to 2 h. Radioactivity in the brain undergoes redistribution in other organs and tissues, as observed in autoradiographic studies. Redistribution is probably due to the extensive breakdown of budipine in the liver, forming the hydrophilic conjugate of metabolite M1. The half-life of redistribution is about 2.5 h. This is about three times more rapid than elimination from the blood. Pharmacologic experiments with rats demonstrate that brain half-life is more relevant for the duration of action than that established in the blood.

Excretion

Excretion of radioactivity is complete (Fig. 5). 40%-50% of the administered activity is on average renally excreted, independent of the route of administration and most of it in the first 24 h fraction. Fecal excretion is consistent with biliary elimination, which is about 50% of total activity within 24 h.

Pharmacokinetics in the Dog

Blood Concentrations and Renal Excretion. The profile of budipine in the blood of dogs (Fig. 6) is similar to that obtained in the rat. Following bolus injection, distribution is pronounced and rapid with a half-life of a few minutes. The elimination phase of radioactivity is independent of the route of administration and can be characterized by a half-life of 23 h, which is about three times longer than that established in the rat.

After oral administration maximum concentration peaks at about 4 h, reflecting half-life of enteral absorption velocity of about 0.6 h as found in other species. In the dog, as in the rat, the major part of the renally eliminated substance is excreted within the first 24 h after administration. One-third of the administered dose is excreted via the kidneys in the dog.

Studies in Humans

Figure 7 shows the serum concentration of budipine after single intravenous injection of 20 mg to healthy volunteers. As can be seen, the serum level curve is characterized again by a double exponential decline. Systemic distribution – the first phase – is very rapid, with a half-life of about 3 min. After having reached 1% of the initial concentration, the terminal phase with a half-life of approximately 31 h covers elimination. Extensive and rapid distribution is reflected also in a large volume of distribution of about 190 l, which is characteristic of budipine.

Following bolus injection of budipine to humans, the concentration of nonconjugated metabolite M 1 in serum, the only one in the human, is below the detection limit (1 ng/ml) of the analytic system. Agreeing with the results obtained in animal experiments, serum concentration is at its maximum at about 4-10 h after oral administration (Fig. 8). Parent compound is predominant, the concentration of the only metabolite M 1 is about 20% of that of budipine at maximum.

On the basis of urine data, absorption rate is nearly 100%, whereas calculation of absolute bioavailability by comparison of area under the curves (AUC) after both administration routes results in a mean value of 47% ($N = 10$). But these findings are typical for budipine as a substance having high affinity to organs or tissue and therefore

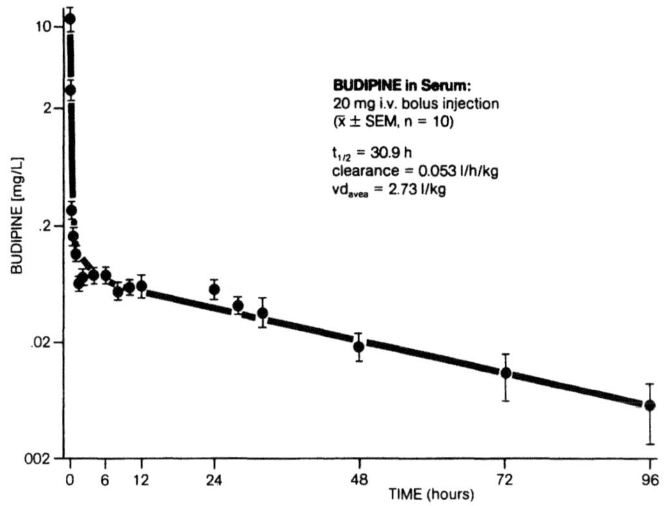

Fig. 7. Serum concentration of budipine in healthy volunteers after a single intravenous dose of 20 mg budipine. The *solid line* represents the concentration-time curve fitted to the experimental data (mean ± standard error, N = 10) according to a two-compartment open body model. Semilogarithmic presentation for illustration of the extensive and rapid distribution

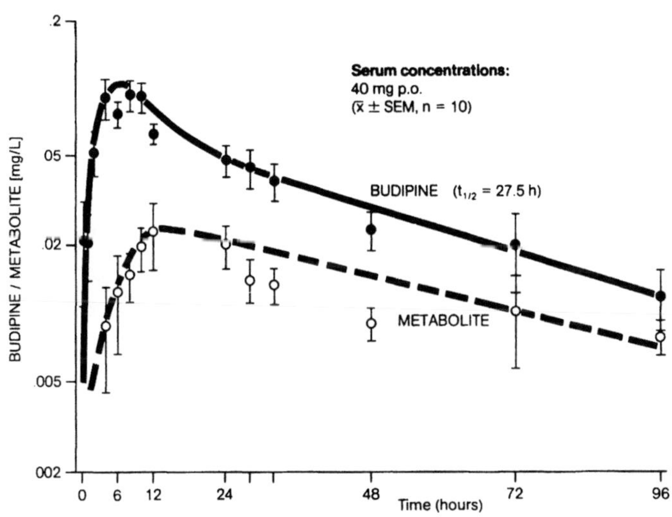

Fig. 8. Serum concentrations of budipine and metabolite M1 after a single oral dose of 10 mg budipine (mean ± standard error, N = 10)

having most of the administered drug not in the central compartment blood. In the human, renal excretion is slightly higher than that in the rat. Approximately 50%-60% of the administered dose is recovered in human urine, 20% as parent compound and 30% as nonconjugated metabolite M1.

Metabolization of Budipine in the Organism

Species dependence in the metabolization of budipine in mouse, rat, dog, and human was studied by comparing the excretion of unchanged budipine and metabolites in the 0 to 24 h urine of animals and 0 to 216 h urine of humans. Hydroxylation of one phenyl residue of the budipine molecule (Fig. 9) is the most important route of biotransformation. While this metabolite M1 represents the only biotransformation product in humans, the conjugated compound is the predominant metabolite in the rat. In the dog, budipine is metabolized to both compounds, conjugated metabolite M1 as well as nonconjugated product. Species dependency in the rate of metabolism and variability in the extent of conjugation explains the species differences in blood half-life (Fig. 10). Extensive metabolism and formation of hydrophilic conjugation product results in a half-life of about 7 h in rat, while breakdown to the nonconjugated metabolite leads to a blood half-life of about 30 h in humans. The appearance of both metabolites in dog, conjugated and nonconjugated M1, results in a blood half-life of 23 h, reflecting an intermediate position between rat and human.

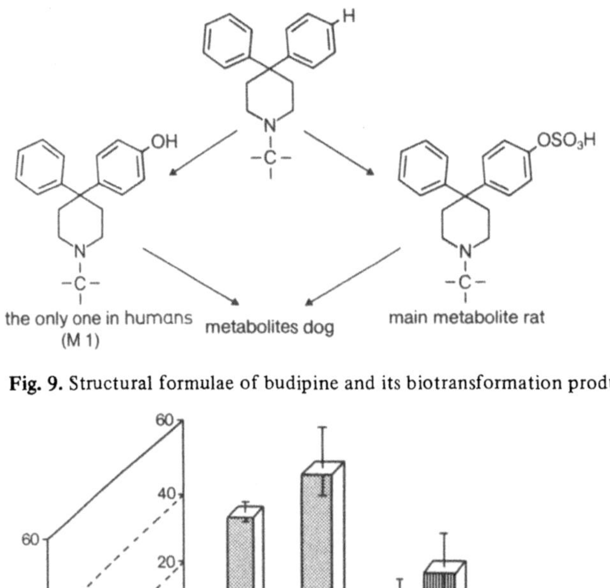

Fig. 9. Structural formulae of budipine and its biotransformation products in human, dog, and rat

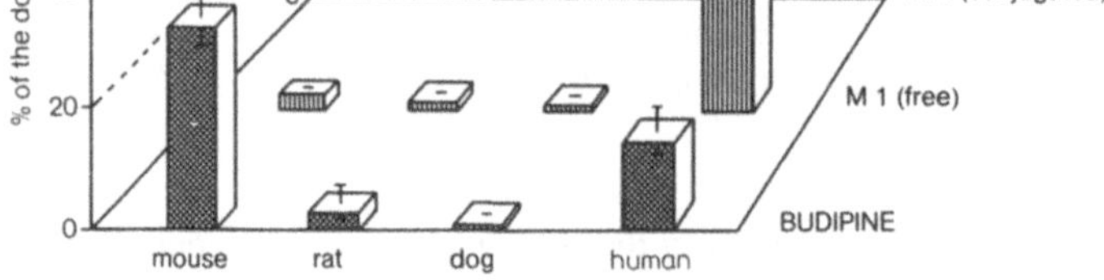

Fig. 10. Excretion of unchanged budipine and metabolites in urine as percentage of the dose after single oral administration of 10 mg/kg to animals and 40 mg to humans. Collection period: for animals 0-24 h, for humans 0-120 h (mean ± standard deviation, N = 10)

References

Dost FH (1968) Grundlagen der Pharmakokinetik. Thieme, Stuttgart

Hackmack G, Menge HG, Eistetter K, Krüger U, Schaefer H, Klosa J (1979) 4,4-Diphenylpiperidine – ein neuer Substanztyp für die Therapie des Morbus Parkinson. Österr Chemie 4: 61

Menge HG, Brand U (1982) Zusammenfassende Darstellung der Pharmakologie von Budipin, einem neuen 4,4-Diphenylpiperidin-Derivat für die Parkinson-Therapie. Arzneimittelforsch 32: 85

Siegfried J, Fischer R (1983) La place pour une nouvelle substance chimique dans le traitement de la maladie de Parkinson et le rôle de la Budipine. Médecine et Hygiène 41: 1977-1981

Clinical Experience with Budipine in Parkinsonian Therapy

W. Poewe, F. Gerstenbrand and G. Ransmayr

Introduction

For about two decades, drug therapy of parkinsonism has centered around oral levodopa substitution (Barbeau 1961; Birkmayer and Hornkiewicz 1961; Gerstenbrand et al. 1963; Cotzias et al. 1967). In recent years, there has been growing awareness of the shortcomings of this form of therapy, characterized by declining clinical efficacy on the one hand and increase in central side effects on the other during long-term levodopa treatment (Fahn and Calne 1978; Gerstenbrand et al. 1978; Rinne 1981). Dopamimetic agonists may frequently serve as an alternative therapy in parkinsonian patients with complications of prolonged levodopa substitution (Calne et al. 1974, 1978; Lieberman et al. 1976), but show declining efficacy themselves (Lees and Stern 1981) and may also induce dyskinesias and paranoid symptoms. For these reasons "classical" antiparkinsonian agents like the anticholinergics and amantadine have regained importance in the treatment of early Parkinson's disease (Fahn and Calne 1978) as well as adjuvants when levodopa dosage has to be reduced because of side effects.

Budipine (BY 701) is a diphenylpiperidine derivative that has shown oxotremorine antagonism as well as antagonism of haloperidol-induced catalepsy and levodopa potentiation in animal experiments (Menge and Brand 1982) which finally led to its clinical testing as an antiparkinsonian agent.

Materials and Methods

A group of 14 parkinsonian patients (6 men and 8 women) with a mean age of 64 years took part in an open clinical trial with budipine; 13 patients had an idiopathic type of the disease, 1 case was of postencephalitic etiology; duration of illness ranged from 1 to 23 years; 12 patients had been pretreated with levodopa, anticholinergics, or amantadine, 2 had been without previous therapy (Table 1). The pretreated patients had all experienced declining efficacy of their drug regimen during the last 2-4 months prior to the study; in three of the L-dopa-treated patients, dyskinesias were present, one had experienced hallucinatory episodes during the preceding 12 months, none of the patients under L-dopa exhibited on-off phenomena.

Before the trial, L-dopa dosage was reduced in the patients having dyskinetic movements to a level where dyskinesias disappeared; in the other patients, the previous antiparkinsonian therapy was kept unchanged and peroral budipine was added in increasing dosage until clinical improvement was observed or a level of 80 mg/day was reached (Table 2). The duration of the trial was 6 weeks.

Table 1. Summary of clinical characteristics of the group of patients treated with budipine

Idiopathic	13	patients
Postencephalitic	1	patient
Mean age (years)	64	(40-72)
Male : female	6:8	
Duration of disease (years)	< 5 = 5 patients	
	5-10 = 4 patients	
	> 10 = 5 patients	
Previous therapy		
L-Dopa	10	patients
Anticholinergics	12	patients
Amantadine	3	patients
None	2	patients

Table 2. Summary of evaluation period, budipine dosage, and concomitant medication in 14 parkinsonian patients

Period of evaluation	6	weeks
Dosage	40-80	mg/day
Concomitant therapy		
L-Dopa (100-600 mg/day)	10	patients
Anticholinergics	5	patients
None	2	patients

The patient's parkinsonian symptoms were rated at weekly intervals using the Webster rating scale, a clinical rating scale with nine items scoring from 0-6 each, and a nurse's rating scale assessing four groups of daily activities (speech, rising from a chair, feeding habits, hygiene). At the end of week 6, a global assessment of overall efficacy of the treatment and of overall drug tolerance was made by the investigator and the patients. Blood pressure and pulse rate were controlled weekly, routine serum parameters were controlled prior to and at the end of the trial.

Results

The mean scores of the items of the Webster rating scale at weeks 0,3, and 6 for the 14 patients are depicted in Fig. 1. A significant decrease in the mean score for the main symptoms: bradykinesia, rigidity, and tremor could be observed both at weeks 3 and 6. Concomitantly there was a significant decrease in the scores for swinging of upper extremities, gait, speech, and self-care, while facies, posture, and seborrhea did not change significantly.

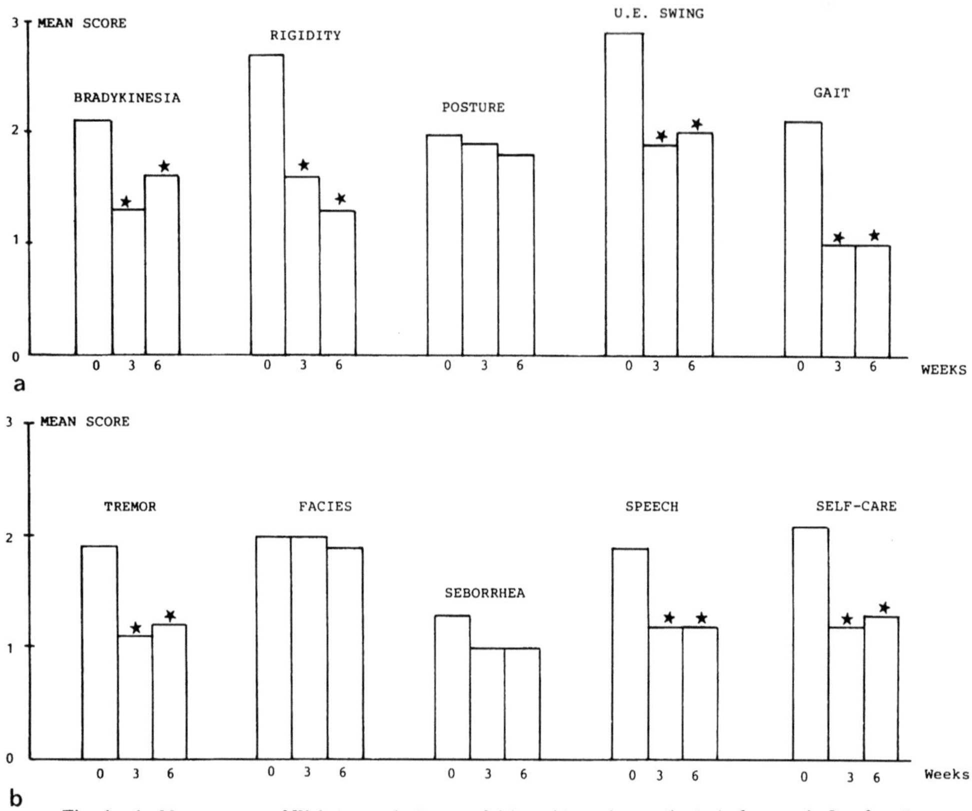

Fig. 1 a, b. Mean scores of Webster scale items of 14 parkinsonian patients before and after 3 and 6 weeks of treatment with budipine. * $P < 0.05$

Similar results were observed with the 7-point clinical rating scale (Fig. 2). However, improvement of tremor did not reach significance whereas facies and posture did improve significantly at week 6 in this scale; also improvement of bradykinesia was significant only at week 6.

When adding the four main bradykinesia-related items of both rating scales, the following efficacy profile on the cardinal symptoms of parkinsonism evolves (Fig. 3): rigidity and bradykinesia improve significantly in both rating systems, while tremor shows significant amelioration only in the Webster rating scale. Vegetative symptoms were not changed significantly during the treatment with budipine.

The nurse's rating for speech, hygiene, feeding habits, and rising from a chair showed improvement of all items from week 0 to 6 which did not, however, reach statistical significance (Fig. 4). Global clinical assessment of overall efficacy and overall tolerance of budipine at the end of the trial ranged from "very good" to "good" in the majority of patients (Table 3).

Four patients experienced side effects with probable relation to the experimental drug during the trial (Table 4). In one patient who had had repeated hallucinatory epi-

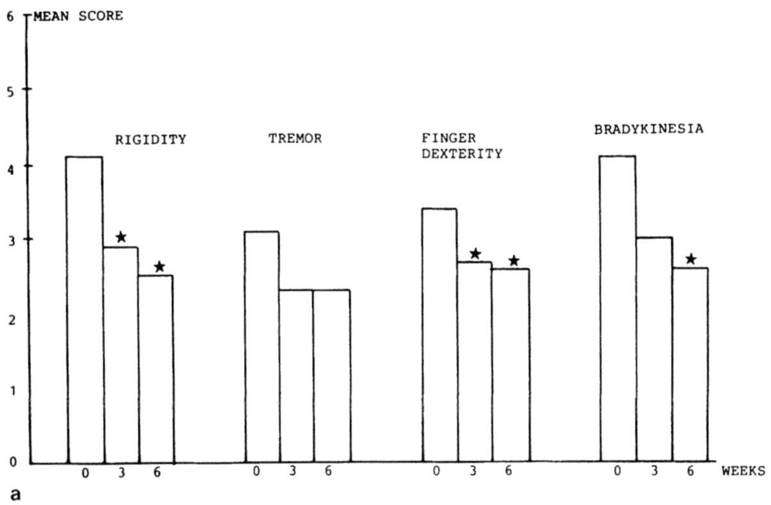

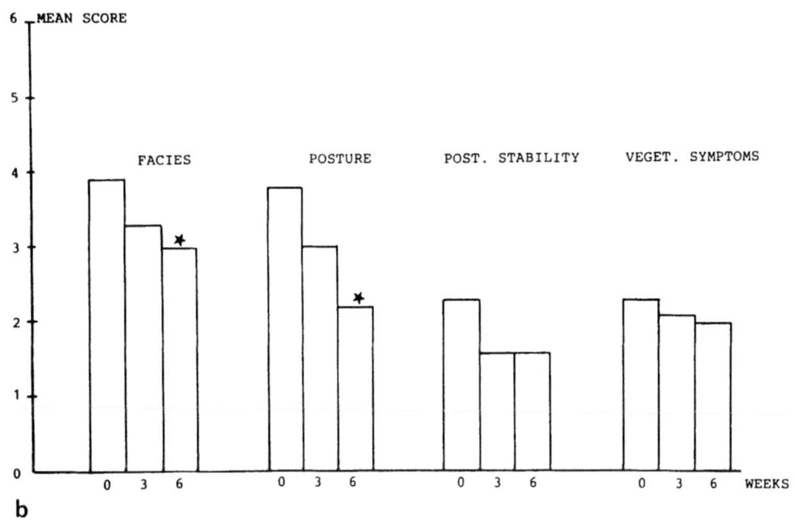

Fig. 2a, b. Mean score of 8 items of 7-point clinical rating scale employed in 14 parkinsonian patients before and after 3 and 6 weeks of budipine treatment. * $P < 0.05$

sodes during L-dopa treatment prior to the trial, budipine treatment was discontinued after 5 weeks because of intermittent disorientation. In another patient, episodes of disorientation at night disappeared despite continuing treatment with budipine. Nausea, headache, nightmares, and sleep disturbances were all reported once and were tolerated or resolved without specific treatment. Two patients discontinued treatment after weeks 3 and 4, respectively, because of lack of efficacy of the test drug.

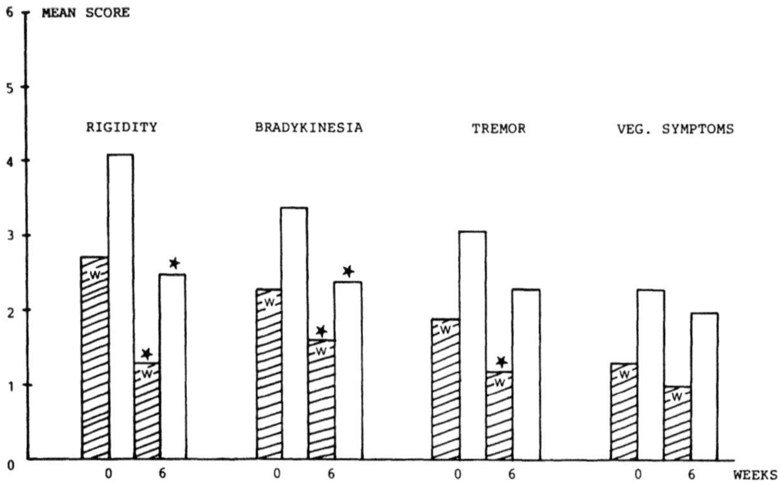

Fig. 3. Comparison of changes of mean scores for major symptoms in both rating scales used in the trial observed between pretreatment phase and after 6 weeks of budipine treatment. * $P < 0.05$

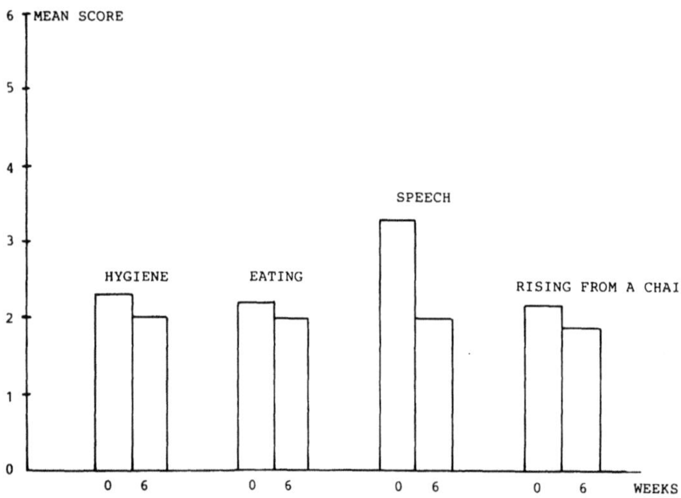

Fig. 4. Change of mean scores of nurse's rating scale for 14 parkinsonian patients from pretreatment phase to week 6 of budipine treatment

Table 3. Global clinical assessment of budipine therapy in 14 parkinsonian patients

	Overall efficacy			Overall tolerance	
	Investigator	Patient		Investigator	Patient
Very good	1	0	Very good	10	6
Good	7	6	Good	3	6
Moderate	3	2	Moderate	1	2
Slight	1	4	Bad		
None	2	2			

Table 4. Summary of side effects observed during treatment with budipine

Frequency	4
Nausea	1
Nightmares	1
Disorientation	2
Headache	1
Sleep disturbances	1

Table 5. Suggested strategy for use of budipine in parkinsonian monotherapy or combination therapy

Early disease phases (monotherapy)
Tremor-dominant Parkinson's disease (adjuvant)
Cases requiring L-dopa dose reduction owing to side effects (adjuvant)
Akinetic crisis with impairment of oral therapy (intravenous therapy)

Discussion

The results of this study demonstrate the clinical effectiveness of budipine (BY 701) in the therapy of Parkinson's disease when used as an adjuvant to L-dopa substitution and/or anticholinergics, but also when used as monotherapy. In this study, bradykinesia and rigidity seemed to be influenced more than tremor, but individual patients with marked tremor did show significant improvement of this symptom. Side effects were few, and generally mild. Caution must be taken, however, in the treatment of patients with prior hallucinatory episodes.

Of 14 patients, 11 completed the trial. Of these 11 patients, 6 remained on continued budipine therapy with L-dopa and/or anticholinergics for more than 18 months with sustained benefit. The two patients on budipine monotherapy, whose duration of illness was less than 2 years at the onset of the trial, received additional L-dopa (300 and 400 mg, respectively) with decarboxylase inhibitor after 6 months of budipine treatment because of progression of the disease. Withdrawal of budipine after the institution of L-dopa in these patients led to deterioration of symptoms so that budipine had to be reinstituted. The main indication for budipine therapy, alone or in combination with L-dopa, that can be derived from our clinical experience with this study and the subsequent (also intravenouse) use of the drug are summarized in Table 5. Budipine thus seems to enrich the spectrum of therapeutic drugs for Parkinson's disease.

References

Barbeau A (1961) Parkinson's disease and parkinsonian syndromes. In: Alema G et al. (eds) Proceedings of the Seventh International Congress on Neurology, Rome. Excerpta Medica, Amsterdam (International congress series no. 38)

Birkmayer W, Hornykiewicz O (1961) Der L-3,4-Dioxyphenylalanin (=L-Dopa) Effekt bei der Parkinson-Akinese. Wien Klin Wochenschr 73: 787

Calne DB, Teychenne PF, Claveria LE, Eastman R, Greenacre JK, Petrie A (1974) Bromocriptine in parkinsonism. Br Med J 4: 442

Calne DB, Plotkin C, Williams AC, Nutt JG, Neiphytides A, Teychenne P (1978) Long-term treatment of parkinsonism with bromocriptine. Lancet I: 735

Cotzias GC, Van Woert MH, Schiffer LM (1967) Aromatic amino acids and modifications of parkinsonism. N Engl J Med 276: 374

Fahn S, Calne DB (1978) Consideration in the management of parkinsonism. Neurology (Minneap) 28: 5

Gerstenbrand F, Pateisky K, Prosenz P (1963) Erfahrungen mit L-Dopa in der Therapie des Parkinsonismus. Psychiat Neurol (Basel) 146: 246

Gerstenbrand F, Poewe W, Rainer J (1978) Neue Entwicklung in der Parkinson-Therapie. Pharmakotherapie 1: 190

Lees AJ, Stern GM (1981) Sustained bromocriptine therapy in previously untreated patients with Parkinson's disease. J Neurol Neurosurg Psychiatry 44: 1020

Liebermann A, Kupersmith M, Estey E, Goldstein M (1976) Modification of the on-off-effect with bromocriptine and lergotril. N Engl J Med 295: 1400

Menge HG, Brand U (1982) Zusammenfassende Darstellung der Pharmakologie von Budipin, einem neuen 4,4-Diphenylpiperidin-Derivat für die Parkinson-Therapie. Arzneimittelforsch 32/1: 85

Rinne UK (1981) Treatment of Parkinson's disease: Problems with a progressing disease. J Neural Transm 51: 161

Controlled Single-Blind Trial of Budipine, L-Dopa, and Amantadine in 102 Patients with Parkinson's Disease

J. Iizuka

Introduction

Information on results of treatment in parkinsonism may be obtained from the patients themselves, from their relatives, i.e., family members and particularly life partners, or their general practitioners. An experienced physician, e.g., consultant neurologist, will draw comparisons with numerous parkinsonian patients in addition to his clinical observations and measurements of the patient in question. His sound and critical judgment, however, can hardly afford to ignore the patient's own feeling, be it ever so transient and subjective. Above all, doctors are supposed primarily to take patients' well-being into consideration, being concerned additionally with the results obtained with sophisticated measurements.

The judgment of positive and/or negative effects of medical agents against parkinsonism may well be severed by a series of factors, e.g., daily alteration of the mental and physical conditions of a patient, since the patient tends to over- or underestimate such factors as a cold, gastric disorder, mental stress, etc., particularly at the initial stage of administration. A similar preoccupation or prejudice against newly developed preparations is not at all rare, even among physicians with isolated experience. This is because physicians and patients are not familiar enough to manage the new medication. There are, then, different viewpoints concerning parkinsonian symptoms between patients and attendants. Anxiety and rigor are primarily felt by patients, while agitation and tremor are obviously noticed by attendants. In other words, the complaints of parkinsonian patients need not correspond to the measurable findings of diverse symptoms on neurological examination. For example, parkinsonian patients suffer profoundly from disturbance in gait following stiffness felt in joints long before tremor develops. They may learn to repress the tremor of a hand, even if it trembles significantly at rest, concentration or by holding it with the other hand, at least at the initial stage of disease. The stiffness felt in arms and legs is hardly noticed by attendants. On the contrary, with the exception of some artists, patients go entirely unaware of gesticulative and even mimic poverty that immediately strike a physician or even a stranger, yet the mimic disturbance may often be accepted at home and later completely ignored by close relatives. Outsiders are inclined to interpret such facial expressions and gesticulations as cold-hearted or even heartless, while the patients are anxious to keep their naturally sensitive emotions repressed. This persistent, almost subconscious repression would lead to hypertonicity of the voluntary muscles in a course of several decades and eventually to depression of all locomotor activity, while the patients feel a vague sense of weakness or exhaustion in the extremities, a state characterized as "shaking palsy" by James Parkinson. There is, certainly, no "palsy" detectable neurologically since the pyramidal tract remains fully intact. Motor fibers of the tract are functionally exhausted following subconscious but persistent stimulation from the brain. When given adrenergic or anticholinergic agents or analeptic drugs in this circumstance, patients usually feel a temporary relief of their generalized weakness.

High doses of analeptics, such as L-dopa, may precipitate patients into an uncontrollable impulse for impulsion, walking the floor restlessly at home or masticating continuously without any food, without having any idea of their bizarre, purposeless, involuntary acts and any memory of them. The patients themselves are hardly aware of these hyperkinetic movements of their extremities or facial muscles, i.e., the symptoms found commonly in chorea, athetosis or automatism. Such performances of nonreflex acts without conscious volition is often interpreted favourably by the patients often as an increase in drive, yet their choreiform or athetoid movements are taken more seriously by their sponses than simple tremor of fingers or rigid facial expression. This is why the judgments of medical effects in parkinsonism differ more or less between patients, their relatives, and attendants and physicians according to the different priorities given to symptoms. Considering these problems in objective evaluation, a primary concern is to compare the effects of budipine with those of two different chemical agents, i.e., L-dopa and amantadine, not just on the basis of quantative measurements, but also by means of qualitative analysis of each of the diverse symptoms in order to assess the therapeutic value of this newly developed drug against parkinsonism in detail.

Patients and Methods

The aim of the present study was to compare the clinical efficacy of budipine with that of L-dopa and amantadine in 102 patients with Parkinson's disease in a controlled single-blind study. The budipine group consisted of 19 men and 15 women with an average age of 64 years (37-75 years) and 63 years (50-70 years), respectively. The L-dopa group was composed of 16 men and 18 women with an average age of 50 years (20-76 years) and 65 years (32-84 years). The amantadine group comprised 12 men and 22 women with an average age of 54 years (20-76 years) and 60 years (34-78 years).

The duration of the disease in the various subgroups is summarized in Table 1. Table 2 gives a summary of the severity of parkinsonism in the different subgroups according to the Webster scale. Concomitant drugs most commonly prescribed were cardiacs and psychotropics. A total of 19 patients in the budipine group, 16 patients in the L-dopa group, and 13 patients in the amantadine group had previously been treated for their condition (Table 3).

The participants in this study were exclusively treated as outpatients, stabilized on the drug, and regularly followed up. No special diet was prescribed. They were requested to take the medication with or after meals with some liquid. Duration of treatment varied between 4 and 176 days. Average dosing in the budipine group, after the stabilization period, was 60 mg/day; the average L-dopa dose in the majority of patients was 250 mg/day; the average amantadine dose was 300 mg/day. Treatment was exclusively oral.

Budipine was provided as a single drug to 19 patients, 15 patients received single or multiple combinations. L-Dopa was given as monotherapy to 11 patients, 23 patients

Table 1. Duration of disease

Period	Budipine *N*	L-Dopa *N*	Amantadine *N*
< 1 year	5	8	21
1-5 years	11	16	7
6-10 years	11	8	5
> 10 years	6	2	1
Not specified	1	–	–

Table 2. Severity of disease

Grade	Budipine *N*	L-Dopa *N*	Amantadine *N*
Mild (0.1-1.0)	–	2	1
Moderately severe (1.1-2.0)	11	18	26
Severe (2.1-3.0)	23	14	7

Table 3. Previous treatment

	Budipine *N*	L-Dopa *N*	Amantadine *N*
Anticholinergics	7	15	9
L-Dopa preparation	6	1	3
L-Dopa + DCI preparation	15	–	2
Amantadine	9	9	–

DCI, Dopadecarboxylase inhibitor

received single or multiple combinations. Amantadine was given as single drug to 25 patients; 9 patients took it exclusively in single combination.

Supportive medication, both as regards type and dosage, was maintained in the budipine and amantadine groups. In the L-dopa group, an anticholinergic was intermediately required. Two patients took benserazide and four patients carbidopa, intermediately in each case. The benserazide and carbidopa dosage was raised in 3 patients each in the course of the study.

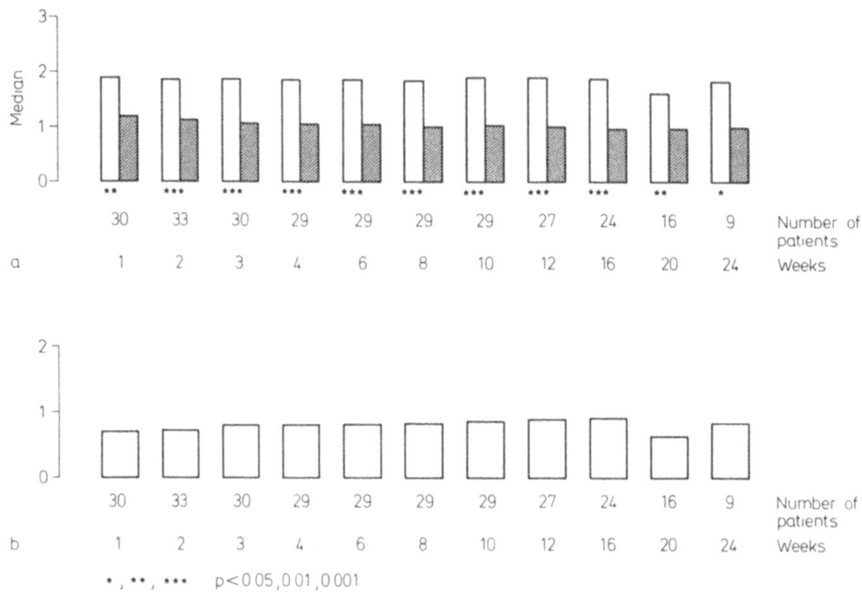

Fig. 1 a, b. Tremor score on the Webster rating scale during budipine treatment **a** median tremor score of the whole group before *(open bars)* and after budipine treatment *(hatched bars)*; **b** mean differences in tremor before and after budipine treatment

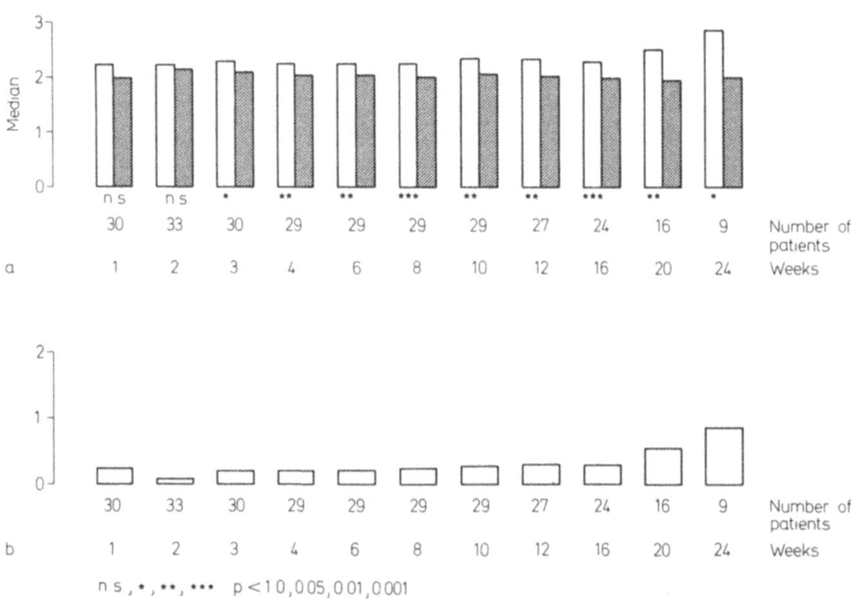

Fig. 2 a, b. Bradykinesia score on the Webster rating scale during budipine treatment **a** median bradykinesia score of the whole group before *(open bars)* and after budipine treatment *(hatched bars)*; **b** mean differences in bradykinesia before and after budipine treatment

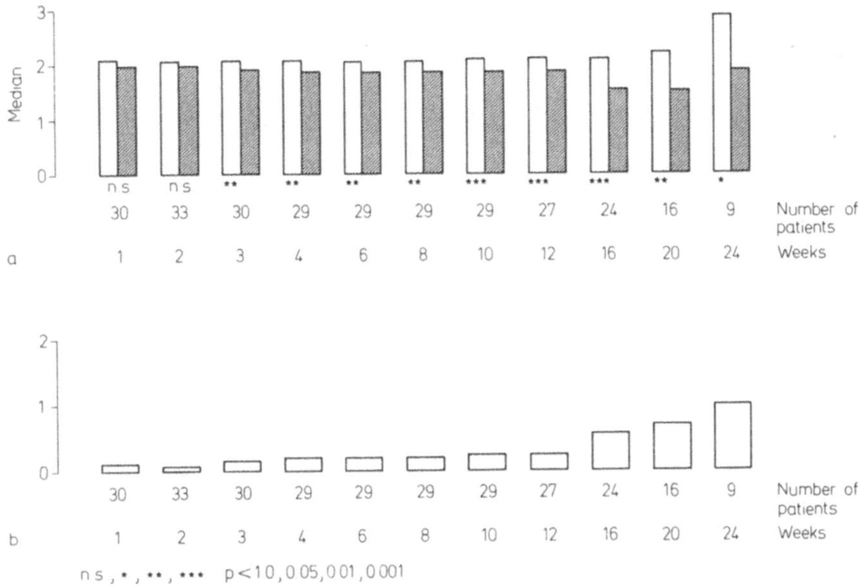

Fig. 3 a, b. Rigidity score on the Webster rating scale during budipine treatment **a** median rigidity score of the whole group before *(open bars)* and after budipine treatment *(hatched bars)*; **b** mean differences in rigidity before and after budipine treatment

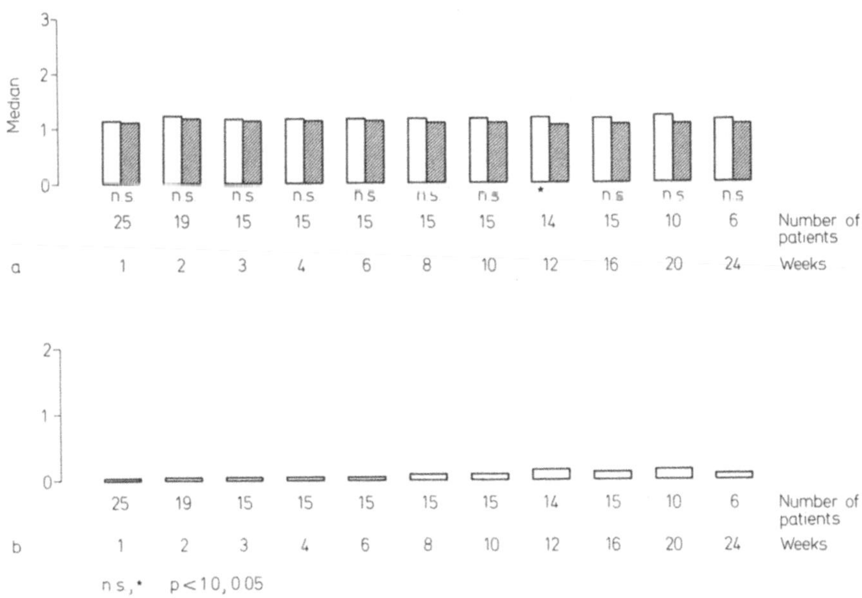

Fig. 4 a, b. Tremor score on the Webster rating scale during L-dopa treatment **a** median tremor score of the whole group before *(open bars)* and after L-dopa treatment *(hatched bars)*; **b** mean differences in tremor before and after L-dopa treatment

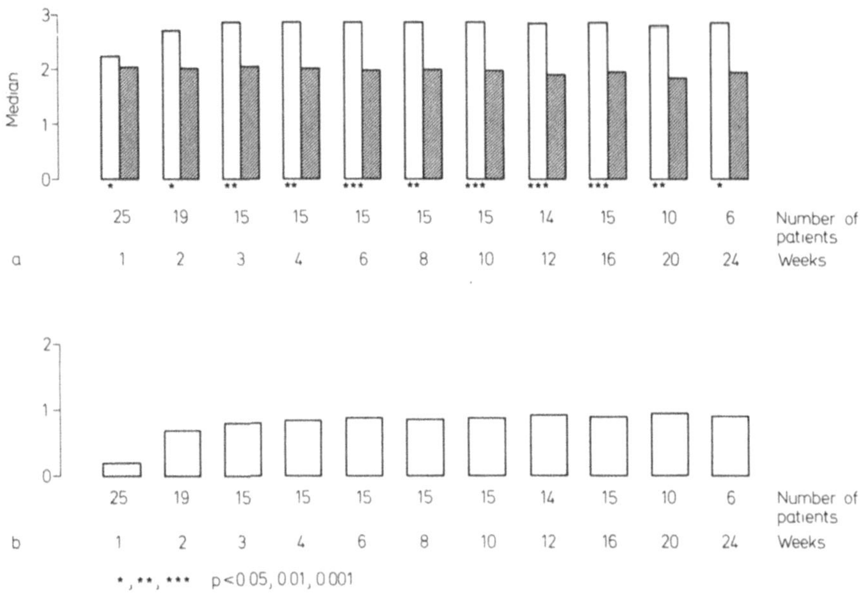

Fig. 5 a, b. Bradykinesia score on the Webster rating scale during L-dopa treatment **a** median bradykinesia score of the whole group before *(open bars)* and after L-dopa treatment *(hatched bars)*; **b** mean differences in bradykinesia before and after L-dopa treatment

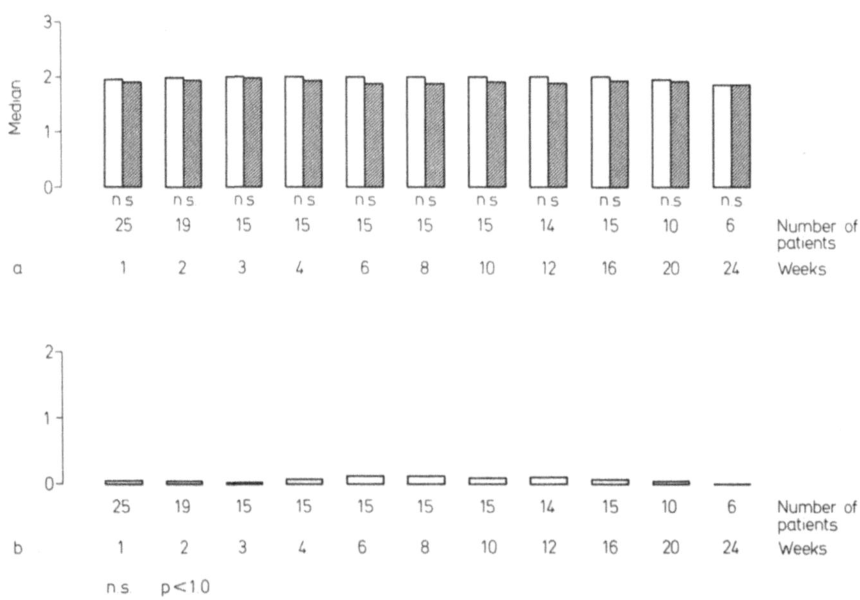

Fig. 6. a, b. Rigidity score on the Webster rating scale during L-dopa treatment **a** median rigidity score of the whole group before *(open bars)* and after L-dopa treatment *(hatched bars)*; **b** mean differences in rigidity before and after L-dopa treatment

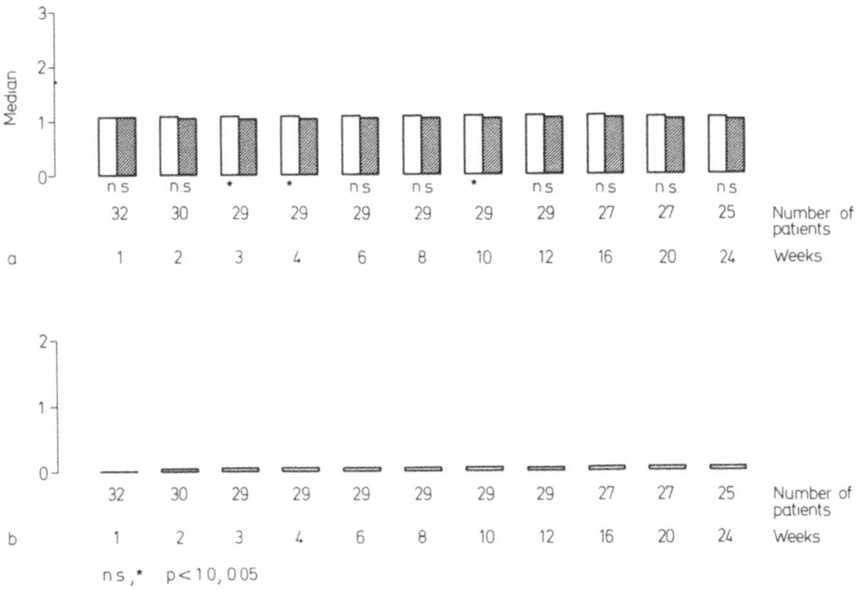

Fig. 7 a, b. Tremor score on the Webster rating scale during amantadine treatment **a** median treatment score of the whole group before *(open bars)* and after amantadine treatment *(hatched bars)*; **b** mean differences in tremor before and after amantadine treatment

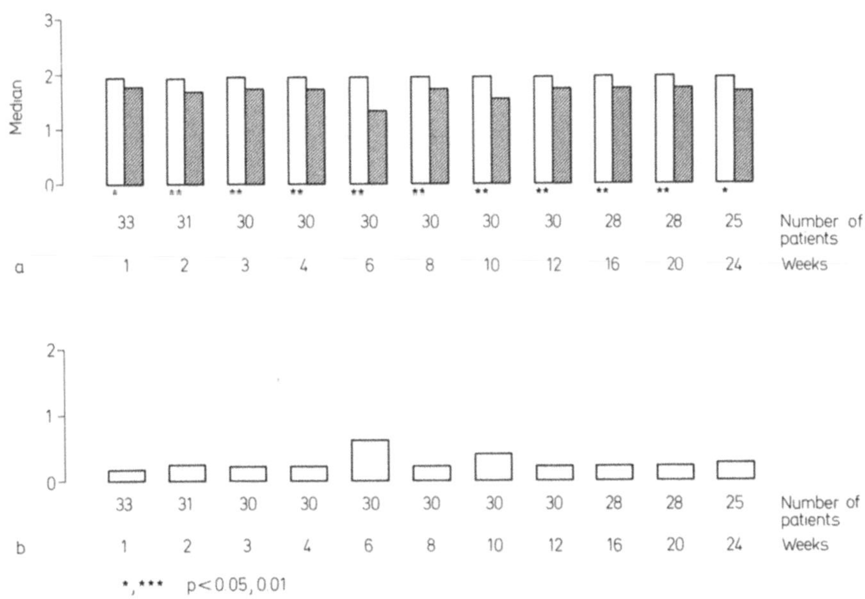

Fig. 8 a, b. Bradykinesia score on the Webster rating scale during amantadine treatment **a** median bradykinesia score of the whole group before *(open bars)* and after amantadine treatment *(hatched bars)*; **b** mean differences in bradykinesia before and after amantadine treatment

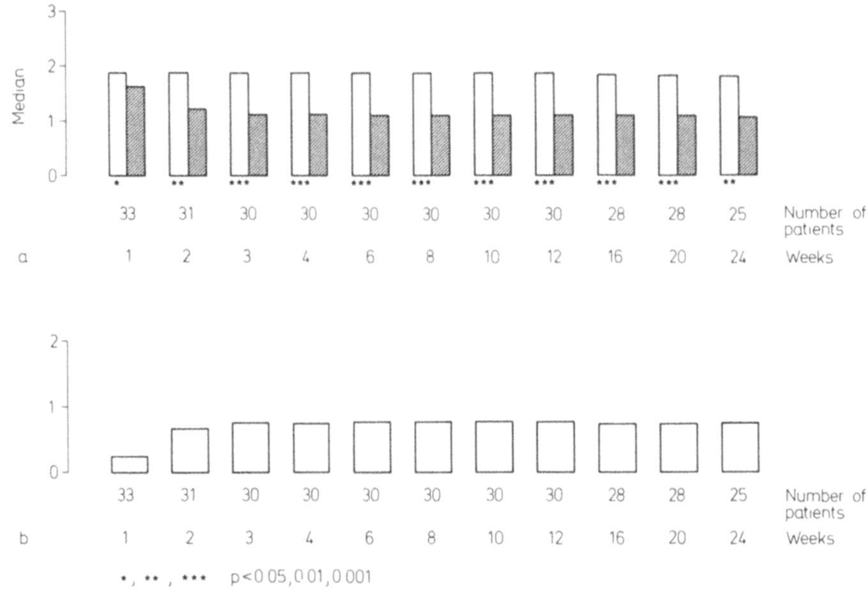

Fig. 9 a, b. Rigidity score on the Webster rating scale during amantadine treatment **a** median rigidity score of the whole group before *(open bars)* and after amantadine treatment *(hatched bars)*; **b** mean differences in rigidity before and after amantadine treatment

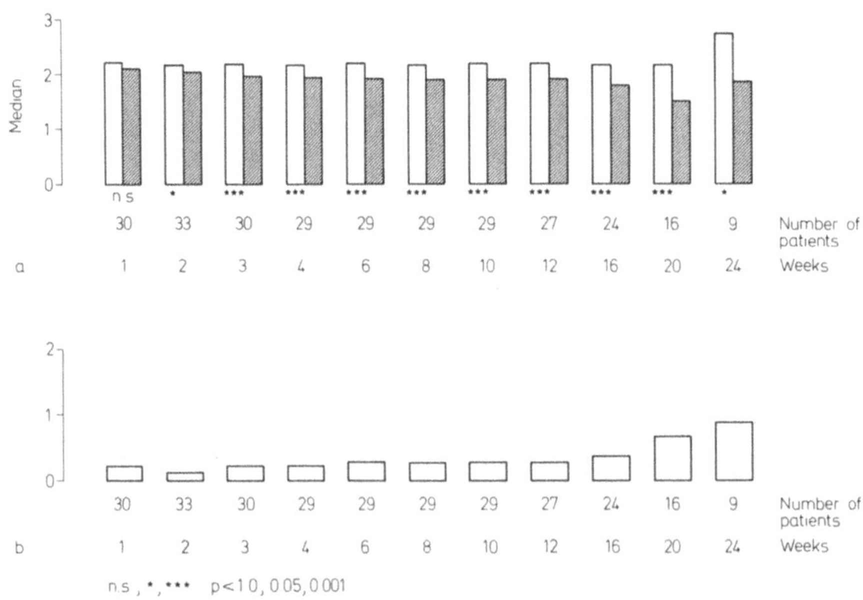

Fig. 10 a, b. Self-care score on the Webster rating scale during budipine treatment **a** median self-care score of the whole group before *(open bars)* and after budipine treatment *(hatched bars)*; **b** mean differences in self-care before and after budipine treatment

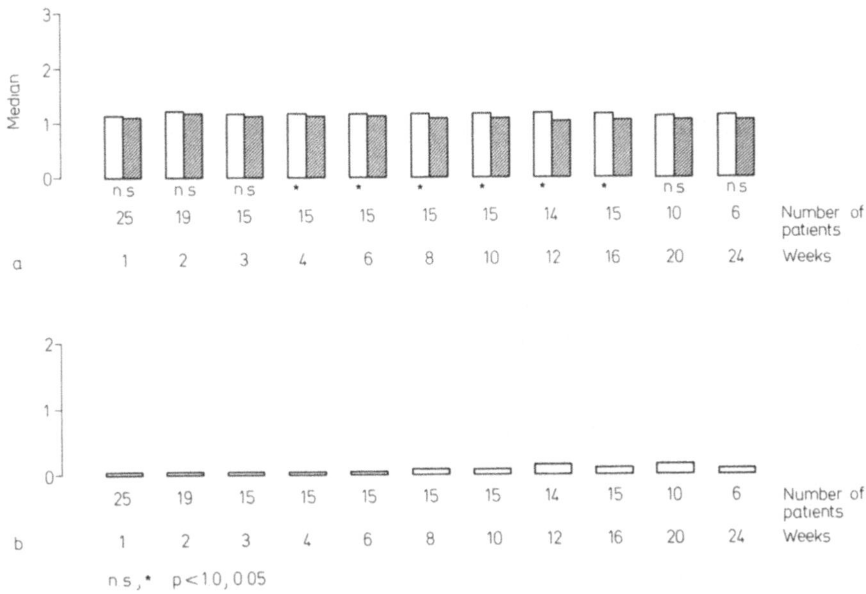

Fig. 11 a, b. Self-care score on the Webster rating scale during L-dopa treatment **a** median self-care score of the whole group before *(open bars)* and after L-dopa treatment *(hatched bars)*; **b** mean differences in self-care before and after L-dopa treatment

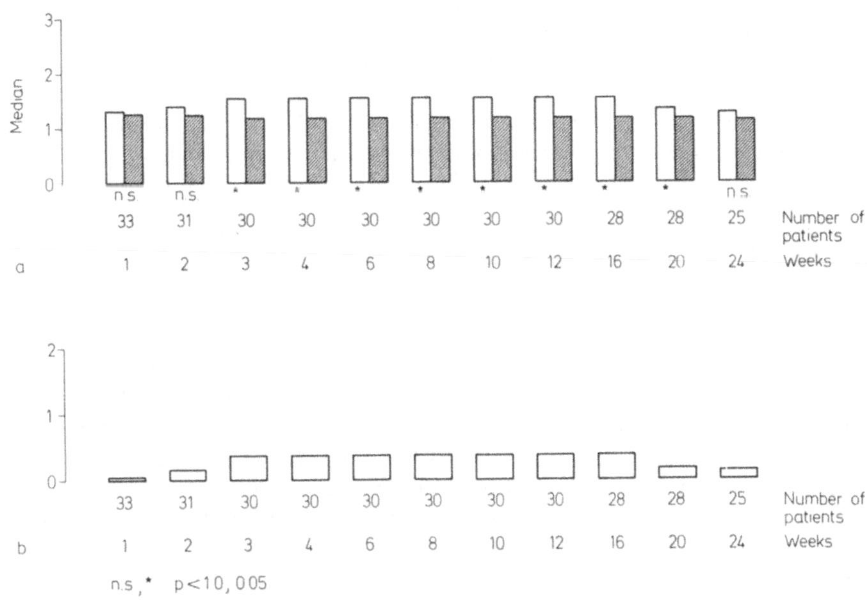

Fig. 12 a, b. Self-care score on the Webster rating scale during amantadine treatment **a** median self-care score of the whole group before *(open bars)* and after amantadine treatment *(hatched bars)*; **b** mean differences in self-care before and after amantadine treatment

Laboratory Values

There were no pathologic changes in laboratory parameters during budipine treatment.

Results

Using Webster's rating scale, tremor, facies, seborrhea, speech, self-care, bradykinesia, rigidity, posture, upper extremity swing, and gait were assessed once a week in the course of treatment. Parkinsonian symptoms improved significantly under all three drugs.

Of the guiding symptoms, tremor improved highly significantly ($P < 0.001$) under budipine (Fig. 1), bradykinesia (Fig. 2) and rigidity (Fig. 3) also improved significantly ($P < 0.05$ to $P < 0.01$) as of the third week. Under L-dopa, only bradykinesia improved significantly ($P < 0.05$) as of the first week (Fig. 4). Rigidity and tremor were not significantly influenced by the treatment (Figs. 5 and 6). Under amantadine, both bradykinesia and, primarily, rigidity improved significantly ($P < 0.05$) in the first week of treatment (Figs. 7 and 8). Tremor improved only slightly (Fig. 9). Of the other parkinsonian symptoms, self-care improved highly significantly under budipine, starting in the second week of treatment (Fig. 1). Under L-dopa and amantadine, improvement was seen in the fourth or third week of treatment, respectively (Figs. 11 and 12).

Thus, budipine succeeded in improving 4 parkinsonian symptoms from an initially moderate grade to a mild one, 16 severe symptoms improved to a moderately severe grade. Under L-dopa, 4 moderately severe Parkinson syndromes improved from a moderately severe to a mild grade, and 3 from a severe to a moderately severe grade. Under amantadine, 9 initially moderately severe parkinsonian symptoms improved to a mild grade and 2 severe symptoms improved to a moderately severe grade.

Side Effects

In the budipine group, 6 patients reported 6 side effects which were questionably related to the drug; 4 patients withdrew from the study within the first 3 weeks. In the L-dopa group, 22 patients reported 34 side effects, of these 10 were definitely drug related; 19 patients discontinued the drug within the first 17 days. In the amantadine group, 16 patients stated 23 side effects of which none was definitely drug-related; 4 patients dropped out between the 5th and 16th day of treatment.

Therapeutic efficacy of budipine was assessed by the investigator as very good to satisfactory in 91% of patients. The same assessment was given by 88% of the patients. For L-dopa, the investigator gave a good to satisfactory assessment in 35%, which was confirmed by 49% of the patients. Therapeutic efficacy of amantadine was assessed by the investigator as very good to satisfactory in 88% of the patients. The same rating was obtained from 81% of the patients.

As for tolerance, the investigator gave a 91% very good to satisfactory rating for budipine which was confirmed by 85% of the patients. Tolerance of L-dopa was objectively rated good to satisfactory in 41% of the patients, but confirmed by only 22% of the patients. The investigator's tolerance rating for amantadine was 91% very good to satisfactory, but only 73% of the patients gave the same rating.

Discussion

Of the guiding symptoms of Parkinson's disease, tremor, rigidity, and bradykinesia were improved by budipine. Amantadine improved primarily rigidity, with bradykinesia next. Tremor was only slightly improved by amantadine, while L-dopa improved bradykinesia only. The fact that self-care was highly significantly improved by budipine deserves to be particularly stressed. Neither amantadine nor L-dopa succeeded in improving this symptom to any essential degree.

All other parkinsonian symptoms improved likewise under budipine (except may be speech). Amantadine had but a slight effect on facies, seborrhea, and speech, while L-dopa had no significant effect on either seborrhea or speech. Improvement of parkinsonian symptoms under budipine was outstanding. Included in this group were 11 patients with an initial moderately severe condition and 23 patients with severe Parkinson's disease. Overall, budipine monotherapy, which was extended to 19 patients, improved 4 moderately severe cases to mild ones and 16 severe states were improved to the moderately severe grade.

Open Clinical Trial of Budipine in 25 Tremor-Dominant Parkinsonian Patients

E. Ketz

Introduction

Budipine was introduced into clinical trials in 1976 and reports have since pointed its efficacy upon parkinsonian symptoms, especially tremor. It is effective in long-term treatment without loss of efficacy and its tolerance is remarkable. The exact neuropharmacologic mode of action of budipine in Parkinson's disease is not yet fully understood. An Oxotremorine antagonism as well as antagonistic effects on reserpine- and haloperidole-induced catalepsy have been shown in animal experiments suggesting a dopaminergic mode (Brand and Menge 1982; see the chapter by Menge and Brand) of action. On the other hand, weak anticholinergic properties have been demonstrated also (see the chapter by Menge and Brand).

Patients and Methods

In the period May 1979 to October 1982, 25 patients (14 women, 11 men) aged 45-87 years for the men and 54-81 years for the women were treated with budipine tablets. The disease pattern was dominated by tremor, the extrapyramidal syndrome being mostly of long-standing duration. All patients were refractory to previous treatment with the major anticholinergic and dopaminergic drugs, with few exceptions. Of the patients, 13.5% had been ill for more than 10 years, 68% for between 1 and 10 years, and 18.5% for 1 year.

More than 60% of the patients had psychopathologic personality changes. Dosage was 3x10 to 4x20 mg budipine, on average, which was probably too low considering the overall length of treatment; 100 mg or more was given in only 5 cases. The duration of treatment was 2-6 months in 18 cases, 5 patients withdrew from the trial for reasons of inadequate effect, 2 patients for inadequate effect and side effects. Monotherapy with budipine was provided in 13 cases, combination therapy in the remaining 12.

For assessment, Webster's rating scale, a scale by the physician, another by the attendants, and self-assessment by the patients, including handwriting, were used. The entire Parkinson's disease complex was checked first once a week and, as of the 10th week, at 4-week intervals. Laboratory routine controls were run every 4 weeks.

Results

According to Webster's rating scale, there was a tendency for bradykinesia and tremor to decrease; the decrease was not significant statistically. Rigidity was uninfluenced.

According to the physician's rating scale, individual symptoms such as tremor and rigidity improved slightly as of the 5th to 6th week of treatment; this was not significant statistically. The parameters of hygiene, eating, speaking, rising from bed, which were assessed according to the attendant's rating scale, were not significantly influenced, except perhaps for hygiene which tended to improve.

Considering individual self-assessments, seven patients believed their condition had improved under budipine, indicating satisfactory antitremor activity. Another five patients gave a good to very good assessment of budipine which coincides with the investigator's and attendant's assessment. Objective side effects of budipine were a pharmacogenic psycho-organic transition syndrome (twice) and urinary retention (once in a patient with a postoperative prostatic condition).

Case Reports

Three cases, shortly summarized here, experienced a rather dramatic amelioration of their tremor.

Case 4, 69-year-old female, right-sided tremor-dominant Parkinson's disease with pronounced psycho-organic syndrome; 4 months 3x20 mg budipine plus Madopar. Tremor improvement was 75%-100%. Apart from slight retching, there were no notable side effects.

Case 11, 56-year-old male, right-sided tremor-dominant Parkinson's disease without essential psychic changes. Combination therapy with Madopar and budipine at a dose of 3-5x10 mg. No side effects, improvement 80%-90%.

Case 25, 75-year-old female, tremor-dominant degenerative Parkinson's disease without asymmetry and without severe psychic changes. Monotherapy with budipine up to 120 mg/day. Marked suppression of tremor up to 90%, reappearing immediately on reducing the dose to less than 100 mg/day. No side effects.

Discussion

Nonoperative antitremor therapy continues to pose problems. In this respect budipine represents a therapeutic enrichment, as documented in the current trial with 25 patients. Even though the 50% success rate is relatively low, the combination of budipine with conventional antiparkinsonian agents, or monotherapy with budipine, may well raise the quality of life of parkinsonian patients, in individual cases essentially. A special advantage of budipine is its weak tendency toward undesirable side effects.

Reference

Menge HG, Brand U (1982) Pharmacological summary of Budipine, a new 4,4-diphenylpiperidine derivative used in treating Parkinson's disease. Arzneimittelforsch 32: 85-98

Effect of Budipine on Parkinsonian Tremor Resistant to Other Antiparkinsonian Medication

G. Ulm

Introduction

Pharmacotherapy of parkinsonian tremor continues to be less than satisfactory, despite the implementation of new therapeutic strategies, including β-receptor blockers. Prodipin (1-isopropyl-4,4-diphenylpiperidine hydrochloride), the precursor of budipine, had already been found superior (unpublished study, Elena Klinik, Kassel, 1977) in this respect in a large number of patients. But the drug had one big disadvantage, it was tolerated on i.v. injection only. Oral doses caused nausea and vomiting and were thus not feasible. We now have high hopes for the product which was further developed, budipine.

Patients and Methods

The study comprises 172 patients who were treated with budipine in addition to other parkinsonian medication. It was carried out over a period of 4 years (from January 1978 to November 1982). The tests were made in open form and were meant to focus the effect of the drug on tremor. The study includes only cases where tremor is the

Table 1. Sex and body weight correlated to body surface (Broca index)

	Patients	
Total	172	♂ 98, ♀ 74
	Mean age (years)	
Total	61	(24-81)
	♂ 59	(37-79)
	♀ 64	(24-81)
	Mean Broca index	
Total	101 %	(76%-147%)
	♂ 102%	(75%-124%)
	♀ 100%	(66%-147%)

Table 2. Concomitant diseases and their medications

Concomitant diseases	Number	Concomitant medication	
Cardiovascular system	116	71	Cardiacs
		8	Coronary agents
		40	Agents enhancing blood flow
Hypertension	16	2	Antiarrhythmics
		1	Vein preparations
		4	Antihypertensives
		3	β-Blockers
		4	Diuretics
Hypotension	4	8	Antihypotensives
Respiratory tract	10		
Urogenital tract	21	7	Urologics
Rheumatic affections	20	4	Analgesics, antirheumatics
		4	Corticoids
Diabetes mellitus	21	14	Antidiabetics
Central nervous disorders	8	1	Antiepileptic
Other	29	29	Unknown

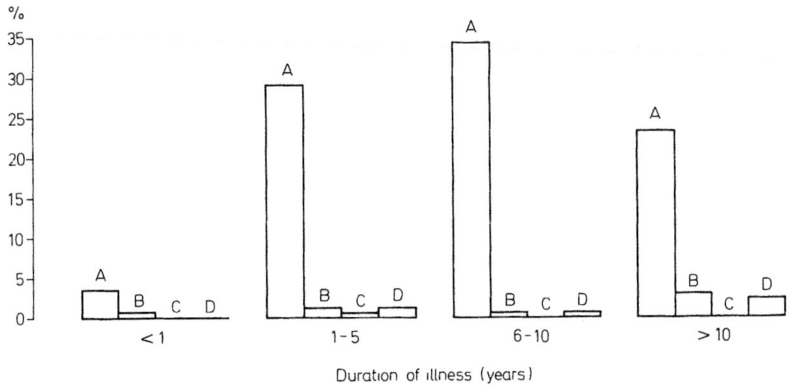

Fig. 1. Diagnosis and duration of illness in 172 patients. *A* endogenous Parkinson's disease plus endogenous Parkinson's disease with vascular implications; *B* postencephalitic Parkinson's disease; *C* posttraumatic and toxic Parkinson's disease; *D* mixed symptoms

primary and most pronounced symptom. These were treated with budipine only when the usual standard methods did not achieve any results or were contraindicated for a variety of reasons. Thus, the conditions under which budipine was tested must be called severe, i.e., where normal methods proved inefficient or unsatisfactory.

On being admitted to our clinic, patients were treated for 10-14 days with combination of various substances to achieve a steady state. During this period, the adjuvant was frequently adjusted to cope with tremor symptoms. If these adjustments proved to be insufficient, we started giving budipine while retaining the adjuvant medication as before. Budipine was first injected, then given orally. The treatment in the clinic lasted between 3 and 5 weeks. After being discharged, the patients were taken care of by their local doctors. The parkinsonian symptoms were measured according to the Webster rating scale before the treatment was started. This was repeated after the injection treatment had been completed (after 10-14 days) and again on the day before the patient was discharged. In addition, the patient was asked to fill in a self-assessment scale, and in many cases a special assistant rating scale was also used. Tables 1 and 2 and Fig. 1 summarize the data for the group of patients in the study.

We started the budipine therapy by giving injections to almost all patients with the exception of a very few, most of whom were obviously afraid. We started with 10 or 20 mg i.v., the latter being the maximum injected dose, which was generally reached by all patients on the 9th or 10th day. We then changed to oral dosages in most cases on the 11th and 12th day. In some cases the injections were continued for up to 14 days. However, most patients were treated orally by that time, the average dose being 60 mg/day. Tentatively, we went as far as 120 mg/day for a limited period, but this

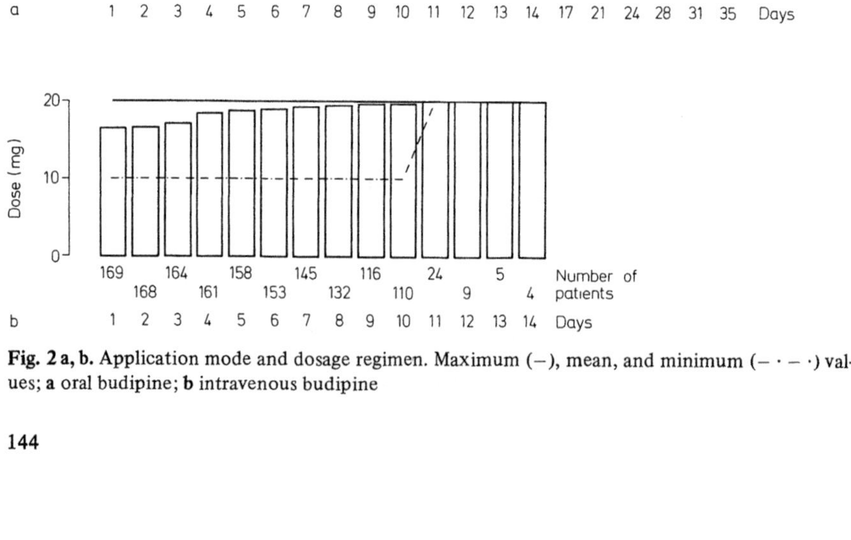

Fig. 2 a, b. Application mode and dosage regimen. Maximum (–), mean, and minimum (– · – ·) values; **a** oral budipine; **b** intravenous budipine

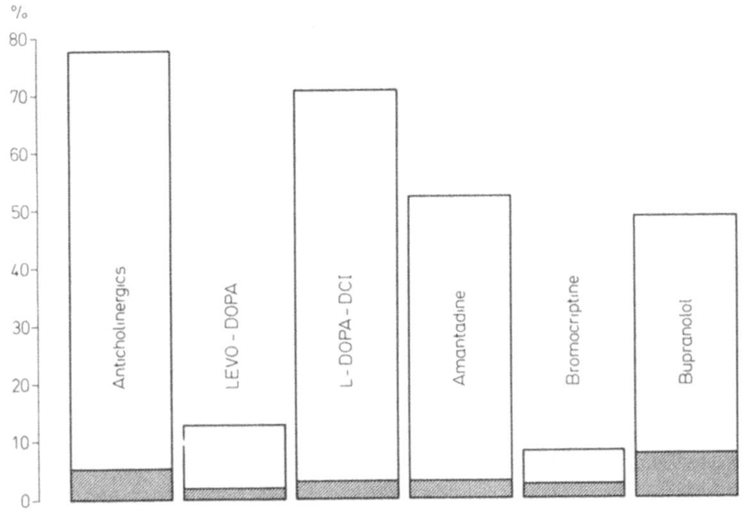

Fig. 3. Concomitant treatment related to number of patients treated. *Hatched area* intermediary only

did not seem to increase the effect. In some cases where budipine had been applied orally right from the beginning, we increased very gradually and individually (Fig. 2).

As we are dealing with tremor-accentuated cases, almost all patients are treated with combination therapy. Anticholinergic medication and β-receptor blockers are widely used. Only in some cases did we have to make slight adjustments during the period in question (Fig. 3).

Results

All parkinsonian symptoms were added up and the sum then divided by the number of symptoms. There is a general improvement, but not a very significant one (Fig. 4). If we isolate tremor, however, we find that the results are indeed much better. Tremor is improved very rapidly, improvement seems to be strongest in week 2.

The considerable improvement in week 6 cannot be confirmed by us and this goes for all diagrams, where week 6 is shown. It is based on the reports of local practitioners, as the patients had already been discharged by that time. One possible explanation would be that the home environment has produced an additional improvement (Fig. 5).

Other symptoms were registered as well. The results show that apart from tremor, there was a marked improvement of bradykinesia. We also noticed a slight improvement of rigidity (Fig. 6). These findings corresponded with an increased self-reliance of our patients (Fig. 7). This independence can also be seen in the assistant rating scale, where personal hygiene and improvements of eating can be observed (Figs. 8 and 9). The patients themselves confirmed this impression. The improvement of tremor and small motor movements was pronounced. Patients felt more independent. There is also a general change of feeling (Fig. 10).

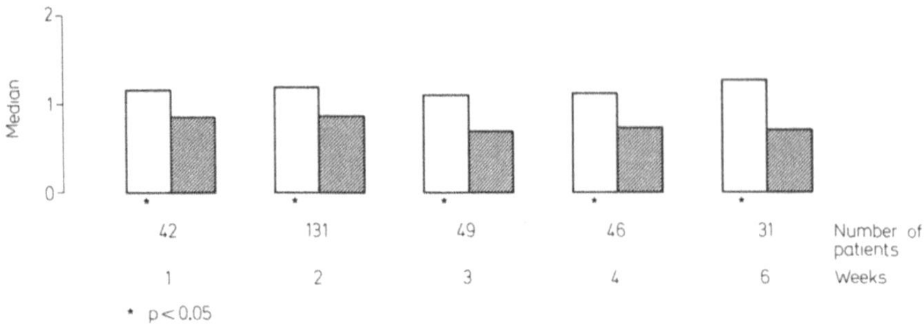

Fig. 4. Therapeutic efficacy, according to the Webster rating scale (sum of all symptoms divided by number of symptoms). Severity: 1 light; 2 moderate; 3 severe

Reasons for discontinuing budipine treatment were either no response to treatment in 16 cases, side effects in 21, and a combination of both in 6 cases. In the 2 remaining cases, the patients became increasingly worried about the fact that a trial preparation was used. In all there were 22 patients who did not respond. In this group, there were 2 patients with essential tremor as well as 3 patients grouped under mixed symptoms. The discontinuations occur relativly early, mostly within the first 10 days. Both desirable and undesirable effects can be observed early. If there is no improvement after the first week, one cannot expect to achieve any success at all (Table 3).

There were 100 side effects in 53 patients during the entire test. By reducing the dosage, some of these could be countered effectively. Among the 45 patients where we had to stop giving budipine, the side effects were responsible for the discontinuation in

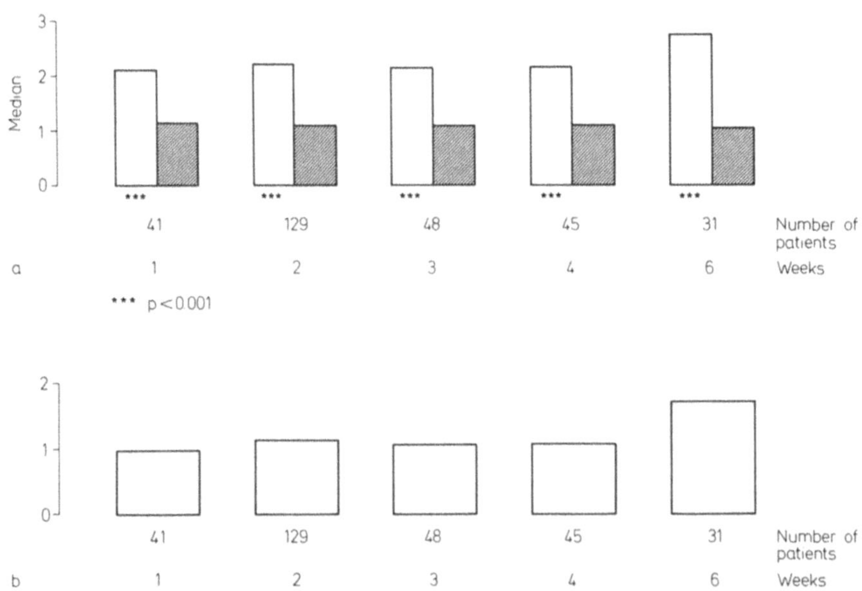

Fig. 5. a Therapeutic efficiency: tremor, according to the Webster rating scale. **b** Differences before and after treatment

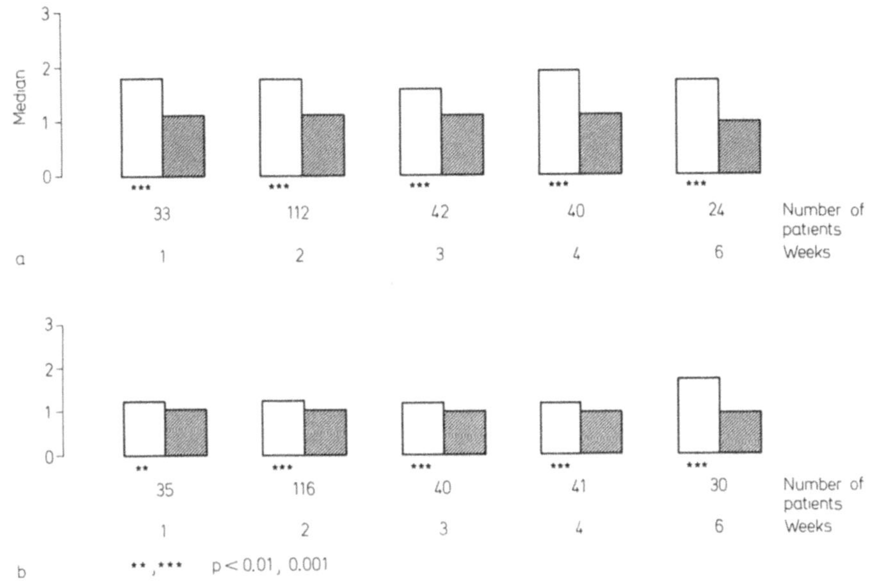

Fig. 6 a, b. Therapeutic efficacy, according to the Webster rating scale **a** bradykinesia; **b** rigidity

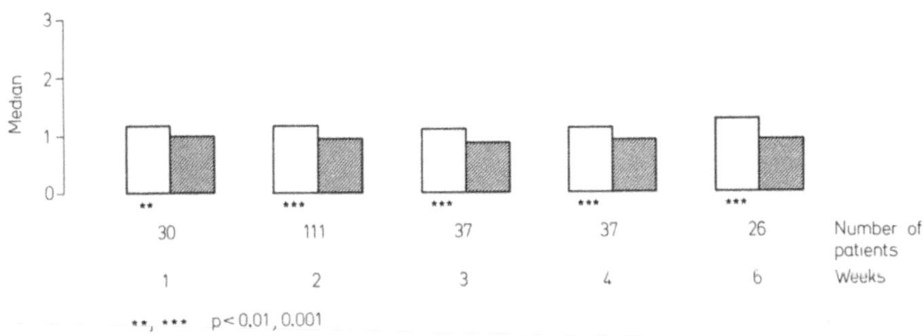

Fig. 7. Therapeutic efficacy: independence, according to the Webster rating scale

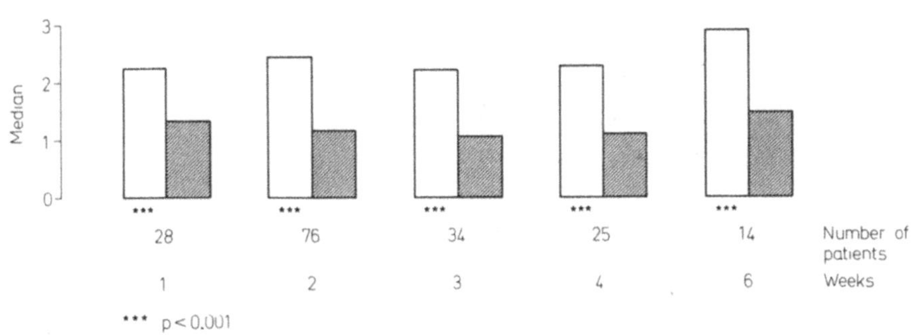

Fig. 8. Therapeutic efficacy: hygiene, according to the assistent rating scale

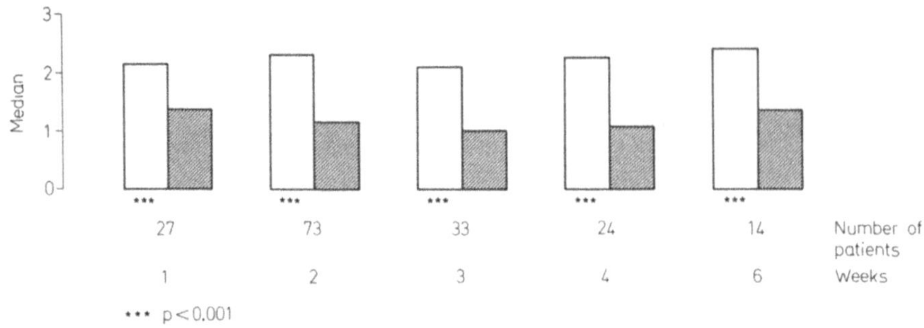

Fig. 9. Therapeutic efficacy: eating, according to the assistant rating scale

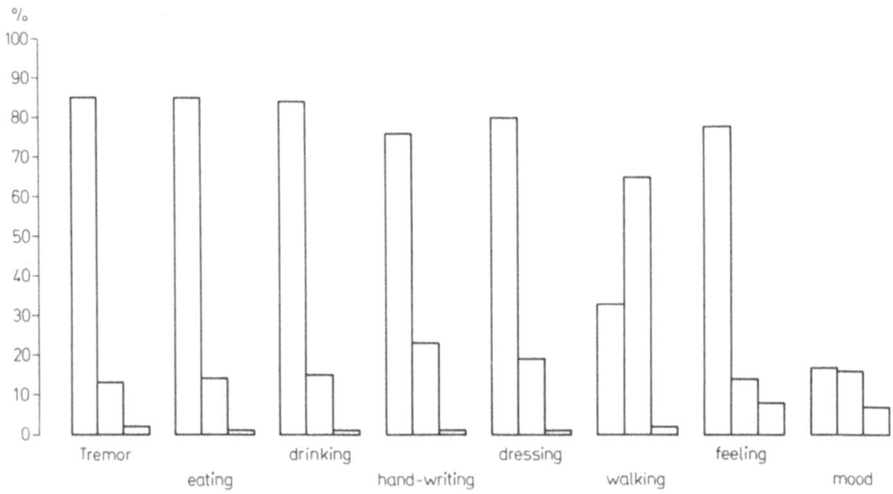

Fig. 10. Influence of budipine on different symptoms, according to self-assessment. *First bar* improved; *second bar* no change; *third bar* deteriorated

Table 3. Reasons for discontinuation of treatment and number of patients who withdrew from treatment

Reason	Number withdrawn[a]
Inadequate effect	16 (14)
Side effects	21 (11)
Inadequate effect and side effects	6 (2)
Nondrug-related reasons	2
Total	45

[a] Figures in parentheses indicate number of patients who withdrew within the first 10 days.

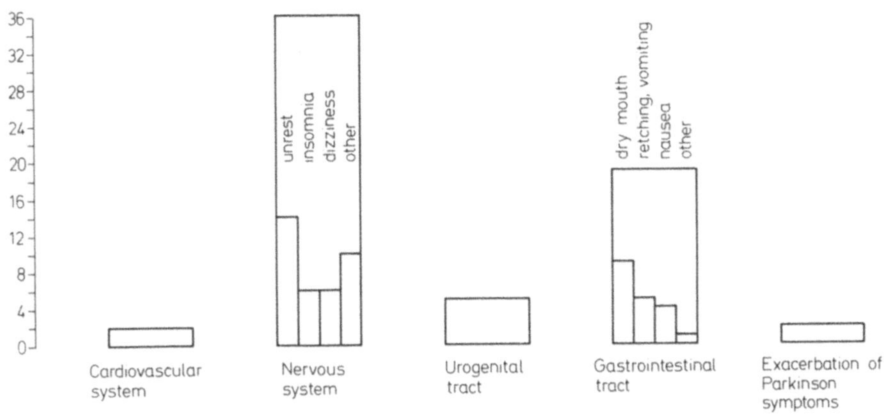

Fig. 11. Side effects related to number of patients. Severity: mild 8%; moderate 48%; severe 4%

only 21 cases. Effects on the central nervous system, e.g., unrest and insomnia, occur most frequently. We also found dizziness and some severe side effects such as the beginnings of thought disorder and disorientation (Fig. 11).

The slight anticholinergic component of the drug can cause urinary retention, grouped here under urogenital tract, but also dry mouth. Other effects were retching, vomiting, and a bitter taste in one instance which is subsumed under "other". Cardiovascular side effects occured in two cases, one where the patient collapsed shortly after the injection and another where the patient complained of a tingling sensation in her legs. The majority of side effects were of moderate severity, only 4% can be called severe.

According to the general assessment of the therapeutic efficacy of the drug by the doctor and the patient, there is no significant difference between the two. The good results stand out very clearly (Fig. 12). The tolerance of the drug is on the whole quite

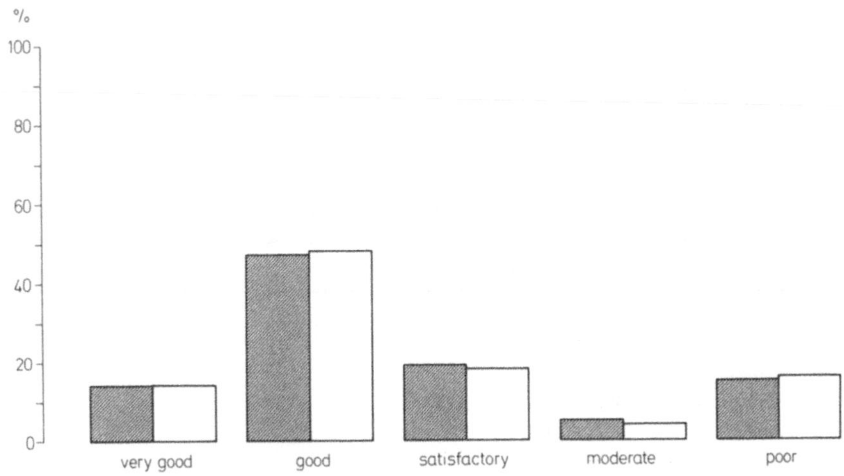

Fig. 12. Therapeutic efficacy related to number of patients (total 165). *Hatched areas* investigator's assessment; *open areas* patients' assessment

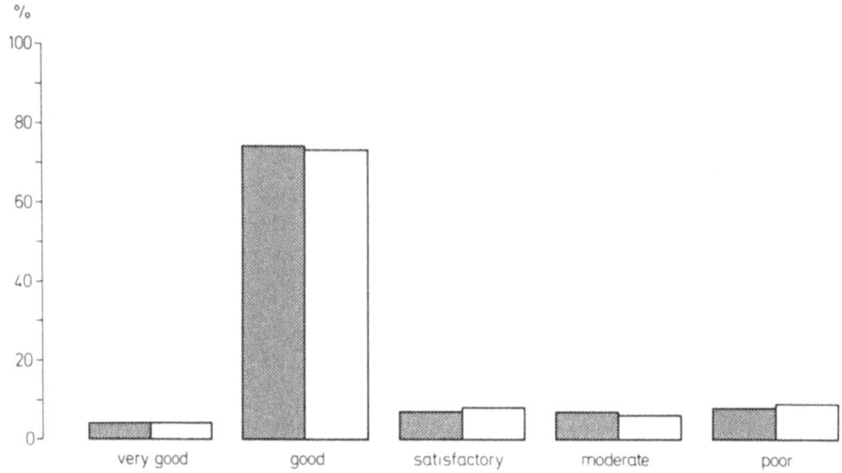

Fig. 13. Tolerance related to number of patients (total 168). *Hatched areas* investigator's assessment; *open areas* patients' assessment

good. Again there is hardly any difference between the doctor's and the patient's assessment (Fig. 13).

Last but not least, we are interested in the question whether budipine can be substituted for other drugs, especially L-dopa preparations. A slight reduction of L-dopa was indeed possible (five patients in all) as well as reductions of anticholinergics and bupranolol. On the other hand, some increases were necessary as well, so that we cannot speak of a general tendency in our sample (Table 4).

Discussion

The present findings clearly document the beneficial effect of budipine on resting tremor. Paradoxically, the compound also has a mood-elevating effect that had not been

Table 4. Dosage regimen of concomitant antiparkinsonian medication during the study

	Dose reduction	Dose increase
Anticholinergics	3	2
Levodopa	1	
Levodopa with decarboxylase inhibitor	4	2
Amantadine		2
Bromocriptine		3
Bupranolol	2	1

readily anticipated since β-receptor blockers or even tranquilizers, which may well have a tremor-depressing effect, were much more liable to cause slight sedation. We could also imagine a dopaminergic effect since levodopa drugs do have an effect on tremor while improving a patient's state of mind. There was no clinical evidence of such a mechanism. Sparing levodopa has not been possible. Dyskinesias were absent. What has been observed is a weak anticholinergic effect, as known for the amantadines. Clinically, budipine ranks somewhere between the anticholinergics and amantadines. Improvements in fine motoricity and rigidity easily fall within this category. Over all, it seems we are dealing with a substance that somehow does not fit any one of the known concepts.

For the practioner, budipine represents an important complement in treating patients with dominant tremor, not in the least because of the good tolerance and low side effect rate. Further studies with monotherapy and long-term therapy should yield important results.

Acknowledgement. I would like to thank the Byk Gulden Company, above all Dr. R. Fischer, for their help in analyzing the data.

Budipine: A New Chemical Substance in the Treatment of Parkinson's Disease*

J. Siegfried and R. Fischer

Introduction

For 15 years now, levodopa has been the most effective drug in the treatment of Parkinson's disease, particularly in combination with a peripheral decarboxylase inhibitor. Administered in the presynaptic area, L-dopa restores the deficient dopaminergic activity by raising intraneuronal dopamine levels, but eventually loses its therapeutic efficacy as a result of progressive destruction of the dopaminergic neurons and hence reduced transformation of L-dopa into dopamine. Temporary insufficiencies of dopamine at the respective brain receptors lead to abnormal movements, oscillations in performance, occasional psychosis, and deficits in mental integrative function. In addition, a dopa-resistant state limits treatment efficacy. Dopamine agonists, i.e., drugs that bypass the degenerating nigrostriatal neurons and directly stimulate the striatal receptors, are considered a valuable alternative. Given in conjunction with L-dopa, the major advantage of these drugs lies in their ability to reduce the on-off effects as the duration of dose action is prolonged. However, the frequently inadequate therapeutic response derives from the fact that it is not possible to give sufficient doses of the drug because of side effects. The search for other drugs is thus desirable, particularly for the treatment of advanced Parkinson's disease, where dopaminergic agents have already been in use for several years. The present 5-year study deals with the clinical effectiveness of budipine against advanced Parkinson's disease, previously treated over a long period of time with L-dopa in combination with a decarboxylase inhibitor, which led to marked side effects, in particular, on-off phenomena.

Methods

Patients

Treated as outpatients for up to 5 years were 13 males and 7 females severely incapacitated by advanced idiopathic Parkinson's disease, who had been under long-term treatment with L-dopa in combination with a peripheral decarboxylase inhibitor (Madopar or Sinemet), had no longer satisfactorily responded to this drug (which was sometimes administered concomitantly with others), and had shown large oscillations in thera-

* This study was carried out with the collaboration of Byk Gulden, Konstanz, Federal Republic of Germany, who supplied the drugs used.

peutic response (on-off phenomena). Informed consent was given by the patients, whose mean age was 58.9 years (37-79 years). The disease was present for 5 years in 4 cases, for up to 10 years in 5 cases, and for more than 10 years in 11 cases. The Webster rating scale was used for neurologic assessment (Webster 1968). The mean score of total disability was 9.7 (6-16). Nine patients were affected by depression. Eight patients underwent a unilateral stereotactic operation (ventrolateral thalamotomy). Two patients had a duodenal ulcer, two were suffering from hypotension, and one fram hypertension. No other relevant concomitant diseases existed. All patients were treated with Madopar from 1 to 11 years (mean duration 6.5 years) except one who received Sinemet. Anticholinergic drugs were taken by nine patients, Symmetrel by three and Parlodel by three, beginning a few days before the start of 5-year budipine treatment. Four patients were also treated with antidepressives. Sixteen patients regularly underwent physiotherapy. Madopar (in one case Sinemet) was given prior to the study in small quantities up to ten times a day with a daily dosage of 375-1250 mg (mean 687.5 mg). Systematic laboratory evaluations were not performed.

One patient only poorly tolerated antiparkinsonian drugs and even showed slight dyskinesia at small doses of Madopar. All others under treatment presented daily changes in performance with abnormal involuntary movements, alternating with freezing episodes, rapid oscillations, and end-of-dose failure (on-off and wearing off effects).

Therapy

At the beginning, the initial daily dose of levodopa (Madopar or Sinemet) and other drugs was maintained and physiotherapy was continued. Budipine was given orally in increasing dosage (up to 3x10 mg within 3-7 days) and later on, depending on clinical observations, 3-4 times a day in higher doses (up to 120 mg in one case; mean dose 47 mg). All patients were summoned for follow-up checks after 1, 4, 8 weeks, and 3 or 4 months. On each visit, every patient underwent a neurologic examination and an interview; the patient's condition was assessed according to the Webster rating scale. In particular, the patient was asked in which period symptom fluctuation was most frequent and overall mobility most impaired; any increase in dyskinesia or the incidence of side effects was also recorded. Statistical analysis was performed using the two-tailed *t*-test at the 5% level of significance. The trial lasted 7 days (see below), or 7-1546 days, respectively (in five patients less than 3 months, in five others between 4 and 6 months, in six between 7 and 12 months, in one up to 18 months, and in three between 3 and 4.5 years; mean duration 368.7 days.

Results

Nineteen patients completed the trial; one developed nausea, vomiting, and dizziness so that treatment was discontinued after 7 days. This case is taken as reference only for the evaluation of side effects.

Table 1. Overall assessment of budipine in the treatment of advanced Parkinson's disease with on-off phenomena after long-term therapy with levodopa and a decarboxylase inhibitor

	Overall assessment			
	Investigator		Patient	
	No.	(%)	No.	(%)
Excellent	2	10.5	1	5.3
Very good	10	52.6	11	57.9
Moderate	3	15.8	2	10.5
Slight	3	15.8	3	15.8
Poor	1	5.3	2	10.5

Table 2. Subjective and objective side effects of budipine at therapeutic dosage

Side effects	Investigator's assessment		Patients' assessment	
	No.	(%)	No.	(%)
None	15	75	12	60
Some minor effects, tolerated	0	0	4	20
Very strong to intolerable	5	25	4	20

Table 3. Frequency of side effects in 20 patients treated with budipine

Nervousness	1	Nausea, vomiting	2
Speech problems	1	Dizziness	3
Impaired vision	1	Confusion	3
Impotence	1	Dryness of mouth	7

Patients' Assessment

The patients' overall assessment is summarized in Table 1. The most valuable benefit for the majority of them was the decrease of daily oscillations in performance. However, tolerance of the drug was assessed differently (Tables 2 and 3). In five cases, the trial was discontinued: after 7 days owing to vomiting (one case), after 29 days owing to dizziness (one case), after 44 days owing to nausea and dizziness (one case), after 53 days owing to nervousness (one case), and after 211 days owing to impaired vision (one case). All except the first of these adverse effects were dose dependent; at lower dosage, however, the benefit of the drug was questionable.

Investigator's Assessment

The investigator's overall assessment is summarized in Table 1; it does not differ significantly from the subjective assessment of the patients. Particular attention was paid to the decrease of the on-off phenomenon without deterioration of performance. This improvement, which could be achieved in 10 of 19 patients after 12 weeks of budipine therapy, was only possible by reducing the daily dosage of Madopar or Sinemet. At the beginning of the trial, the mean daily dosage of L-dopa was 684 mg. By the last control day, it had decreased to 527 mg. For quantitative evaluation, again the Webster rating scale was used. The mean value of this scale amounted to 9.6 at the beginning of the trial. On the last control day, a mean value of 8.6 was assessed, thus, the scale was improved in 8 of 19 patients and there was no deterioration. This overall amelioration was observed in the first few days after optimum dosage of budipine and did not change again throughout the trial. Symptoms that improved were bradykinesia in one case, rigidity in five cases, posture in two cases, arm-swing in four cases, gait in two cases, tremor in one case, facies in one case, speech in two cases, and self-care in four cases. Increasing incidence of tremor was noticed in two cases. Improvement of the depressive condition was reported in five of nine cases. The side effects are summarized in Tables 2 and 3.

Discussion

Fluctuation in response to L-dopa is a common and serious complication of long-term therapy. It might be due to further degeneration of both dopaminergic and noradrenergic neurons as the disease progresses, thus resulting in a loss of the ability to store dopamine and to synthesize noradrenaline from L-dopa (Marsden et al. 1981). A large number of different drug-induced abnormal movements can be observed (Barbeau et al. 1970) and these dyskinesias are the first clinical manifestation of levodopa overdosage; they soon alternate with lack of effect of levodopa (underdosage). This oscillations in performance, appearing in the majority of cases after 24-36 months of therapy (Barbeau 1972; Lee et al. 1971; Markham 1972), can be described in various terms: end-of-dose akinesia, on-off phemomenon, akinesia paradoxica (Barbeau 1976), peak-dose akinesia, peak-dose dyskinesia, phasic dyskinesia, paroxysmal dyskinesia, peak-dose delirium (Marsden 1980), early-morning akinesia, nocturnal akinesia, daily occurrence, rapid oscillations, freezing episodes (Rinne 1980), etc. The simplest expression to cover all these manifestations is on-off phenomena. One of the first approaches to control this major complication of treatment consists in reducing each levodopa dose (but not the daily dosage) by increasing the frequency of drug (Marsden 1976; Rondot 1981; J. Siegfried 1979, unpublished) intake. Temporary drug withdrawal (drug holiday) has been advocated (Weiner et al. 1980), but is not appropriate because of serious consequences such as dysphagia, aspiration pneumonia, and eventually death. It has been assumed that degeneration of the nigrostriatal system progresses to the point at which the remaining neurons are incapable of synthesizing dopamine from levodopa in sufficient quantities (Liebermann et al. 1976). Dopaminergic transmission can be restored by drugs that bypass the degenerating nigrostriatal neurons and directly stimulate the dopaminergic receptor in the postsynaptic area. Agonists of the dopaminergic receptor are apparently capable of acting upon the postsynaptic dopaminergic receptors, which are preserved even after the afferent dopaminergic nerve endings have disappeared. Apomorphine and its derivatives, piribedil, bromocriptine, and other derivatives of rye ergot such as ergotrile, lisuride, and pergolide, are the most widely used dopaminergic receptor agonists. However, their effect is limited, despite satisfactoy improvement in some patients.

Budipine (1-*tert*-butyl-4,4-diphenyl-piperidine hydrochloride) has provided pharmacologic and biochemical data indicating that the drug has some indirect stimulatory effect on dopaminergic transmission and anticholinergic properties (Menge 1978). This drug produces contractions in the isolated rat uterus and can therefore be classified as a direct acting serotonergic agonist (Eltze 1980). It has also been demonstrated that budipine interferes with the uptake and storage of serotonin in human blood platelets, wich are used as a simple analogous model of serotonergic nerve terminals (Eltze 1980); in vitro, this drug has a dual effect: it inhibits uptake and induces the release of previously absorbed serotonin. A decrease of serotonin in the brain of patients with Parkinson's disease has been proved for nucleus caudatus, putamen, pallidum, thalamus, hypothalamus, and substantia nigra (Bernheimer et al. 1961; Hornykiewicz 1966). Using radiolabeled ligand techniques, it has also been shown that specific binding of serotonin is significantly lower in the putamen of individuals suffering from Parkinson's disease, reflecting a decrease in the number of serotonergic receptors of about 37% (Reisine et al. 1977). Furthermore, it has been assumed that budipine induces an increase in the activity of postsynaptic serotonergic fibers as a result of: (a) direct serotonergic activity at postsynaptic receptor sites; (b) presynaptic inhibition of serotonin uptake; (c) induction of stored amine release; and, to a lesser extent, (d) inhibition of monoamine oxidase (Eltze 1980). In parkinsonian patients treated with budipine, an increase in amino acids in serum and CSF has been demonstrated (Marsden 1980).

Clinically, the antiparkinsonian effect of budipine has been documented (Przuntek 1979; Ulm 1981) and its favorable influence on tremor has been emphasized (Brand et al. 1979, 1980; Przuntek 1978). However, budipine seems to exert its most favorable effect when combined with levodapa; in this way, the levodapa dosage can be reduced, which in turn leads to a decrease in the intensity of the on-off phenomena (Ulm 1981). It can be clinically assumed that budipine promotes the storage of L-dopa or acts as a potent, long-term central dopaminergic agonist. The treatment of advanced Parkinson's disease remains a considerable challenge. Levodopa is still the best remedy, particularly in the early stage of the disease, but later combination with other substances is indispensable. In this respect, budipine appears most promising.

References

Barbeau A (1972) Long-term appraisal of levodopa therapy. Neurology (Minneap) 22: 22-24

Barbeau A (1976) Pathophysiology of the oscillations in performance after long-term therapy with L-dopa. In: Birkmayer W, Hornykiewicz O (eds) Advances in parkinsonism. Roche, Basel, pp 424-434

Barbeau A, Mars H, Gillo-Joffroy L, Arsenault A (1970) A proposed classification of dopa-induced dyskinesias. In: Barbeau A, McDowell FH (1972) (eds) L-dopa and parkinsonism. Davis, Philadelphia, pp 118-121

Bernheimer H, Birkmayer W, Hornykiewicz O (1961) Verteilung des 5-Hydroxytryptamins (Serotonin) im Gehirn des Menschen und sein Verhalten bei Patienten mit Parkinson–Syndrom. Klin Wochenschr 39: 1056-1059

Brand U, Menge HG (1980) Antagonismus gegen den Tremor nach N-Carbamyl-2-(2.6-dichlorophenyl)-acctamidin-hydrochlorid (LON-954) und Oxotremorin. Eine vergleichende Studie zum Wirkungsmechanismus. Drug Res 30: 1242-1243

Brandt F, Oppel F (1979) Quantitative measurement of parkinsonian tremor before and after stereotactic operation. Adv Neurosurg 7: 197-202

Eltze M (1980) The effects of prodipine and budipine on 14 L-5-hydroxytryptamine uptake and release by human blood platelets. Drug Res 30: 1129-1134

Gründig E, Gerstenbrand F (1980) Vergleichende Untersuchungen über die Wirkung von L-Dopa, Prodipin und Amantadin auf den Aminosäurespiegel von Serum and Liquor beim Parkinson-Syndrom. Wien Klin Wochenschr 42: 868-871

Hornykiewicz O (1966) Dopamin (3-hydroxytyramine) and brain function. Pharmacol Rev 18: 925-964

Lee JE, Sweet RD, McDowell FH (1971) Treatment of parkinsonism with levodopa: follow-up after two years of treatment. Ann Intern Med 75: 703-708

Liebermann AN, Kupersmith M, Estey E, Goldstein M (1976) Treatment of Parkinson's disease with bromocriptine. N Engl J Med 295: 1400-1404

Lücking CH (1978) Medikamentöse Therapie des Parkinson-Syndroms. Pharmakotherapie 1: 182-189

Markham CH (1972) Thirty months trial of L-dopa. Neurology 22: 17-21

Marsden CD (1976) Advances in the management of Parkinson's disease. Scott Med J 21: 139-148

Marsden CD (1980) On-off phenomena in Parkinson's disease. In: Rinne UK, Klingler M, Stamm G (eds) Parkinson's disease – Current progress, problems and management. Elsevier/North Holland, Amsterdam, pp 241 251

Marsden CD, Jenner P (1981) L-dopa's action in Parkinson's disease. Trends Neurosci 4: 148-149

Menge HG (1978) Pharmacodynamics of two substances belonging to a new chemical class with antiparkinson activity. 11th C.J.N.P. Congress, Vienna, p 53

Przuntek H (1978) Medikamentöse Behandlungsmöglichkeiten bei verschiedenen Tremor- und Myoklonieformen. Pharmakotherapie 1: 195: 199

Przuntek H (1979) Therapie der Parkinsonschen Krankheit. Med Klin 74: 1077-1083

Reisine TD, Fields JZ, Yamamura HI, Bird ED, Spokes E, Schreiner PS, Enna SJ (1977) Neurotransmitter receptor alterations in Parkinson's disease. Life Sci 21: 335-344

Rinne UK (1980) Recent advances in the treatment of Parkinson's disease. Acta Neurol Scand (Suppl) 78: 139-148

Rondot P (1981) Intérêt d'un dosage faible de l'association levodopa-bensérazide dans le traitement de la maladie de Parkinson. Med Hyg 39: 3832-3837

Siegfried J (1978, unpublished) Comparison of some new drugs in the management of dyskinesias. 6th International Symposium on Parkinson's disease, Quebec 1978

Siegfried J (1979) Acquisitions en neurologie-neurochirurgie. Med Hyg 37: 33-34

Ulm G (1981) Aktuelle Therapie beim Morbus Parkinson. Kassenarzt 21: 5231-5234

Webster DD (1968) Clinical analysis of the disability in Parkinson's disease. Mod Treatment 5: 257-282

Weiner WJ, Koller WC, Perlik S, Nausieda PA, Klawans HL (1980) Drug holiday and management of Parkinson disease. Neurology 30: 1257-1261

Observations on the Therapeutic Effect of Budipine on Parkinson's Disease

Wang Xinde

Introduction

Budipine is a new antiparkinsonian drug which has been shown to ameliorate the major parkinsonian symptoms such as tremor, akinesia, and rigidity. It may have some influence on dopamine release and uptake and may also possess anticholinergic properties and exert definite stimulatory effects on the serotonergic receptors (Menge and Brand 1982).

Patients and Methods

We performed a clinical trial of budipine in 16 cases of Parkinson's disease. There were 10 patients with rigidity and akinesia and 4 with tremor as the predominant symptoms, and 2 with on-off phenomena and dyskinesia during the course of treatment with levodopa. Among the 16 patients, there were 12 males and 4 females. The oldest patient was 72 years and the youngest 48 years. Of the 16 patients, 13 were treated by oral administration of budipine. Therapy was begun with 5 mg p.o. on the first and second day. As of the third day, the daily dose was increased by 5 mg until a therapeutic effect occurred. The maximal daily dose was 30 mg. Another three patients (two with akinesia and rigidity, one with tremor) were treated by intravenous or intramuscular injection of 20 mg budipine. ECG, blood analysis (including hemoglobin and platelet count), urinalysis, serum potassium, sodium, and chlorides, serum alkaline phosphatase, fasting glucose and transaminase, and liver function tests were done before and after the treatment. Blood pressure was measured to see whether there was postural hypotension or not. Scoring was done according to the Webster rating scale to note the improvement of symptoms and signs. Side effects of budipine were closely followed during the course of treatment.

Results

Therapeutic results for the 13 patients treated by oral administration of budipine were as follows. In eight patients, there was marked amelioration of degree of rigidity, improvement of akinesia, and increase in activity after treatment with budipine. The re-

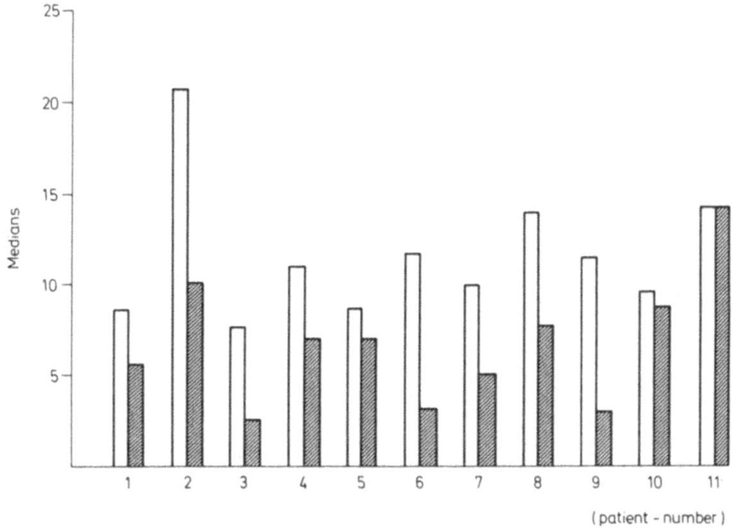

Fig. 1. Therapeutic results of budipine treatment in 11 of 16 treated parkinsonian patients as judged by the Webster rating scale (*open bars* pretreatment score; *hatched bars* posttreatment scores)

duction in Webster score before and after treatment was in the range 2-11 with an average of 6.3 (Fig. 1). Among three cases of Parkinson's disease with tremor as the predominant symptom, there were two cases in which the tremor was markedly ameliorated after treatment with budipine. The Webster score in these two cases was reduced by 2 and 5 points, respectively; the reduction is due mainly to amelioration of tremor. Another case with tremor as the predominant symptom, which was resistant to levodopa therapy, experienced no therapeutic effect of budipine therapy.

In the two patients with on-off phenomenon and dyskinesia after treatment with levodopa, one had marked on-off phenomenon before budipine therapy. At the off time, the patient appeared rigid and could not move at all; the score given then was 25-27. At the on time, however, he was dyskinetic, with increased movement of both upper extremities, head, and face; the score was reduced to 3-5, the patient had several on-off periods every day, the off duration was much longer than the on duration. When he was treated with budipine, 10 mg/day the on-off phenomenon was markedly alleviated and dyskinesia also disappeared. The Webster score taken at different times of the day was kept between 5 and 15 (Fig. 2). Another patient who began to take Madopar 5 × 250 mg in January 1981 and later on shifted to levodopa 750 mg, developed on-off phenomena and dyskinesia after 1 month. After administration of budipine 20 mg/day, on-off phenomena and dyskinesia were reduced by 50% after 2 weeks. The dosage of budipine was later increased to 30 mg/day and these symptoms were further reduced by 70%. In these two cases, budipine was given for 1.5 years and still has a good effect.

In this series of 16 patients, 11 were treated with maximal dosage of 30 mg budipine; the maximal dosage in the remaining 2 patients was 20 and 50 mg, respectively. Three patients were treated intravenously or intramuscularly with a dosage of 20 mg. There was marked improvement of akinesia, rigidity, and tremor. Each injection of budipine could maintain the therapeutic effect for 24-48 h (Table 1).

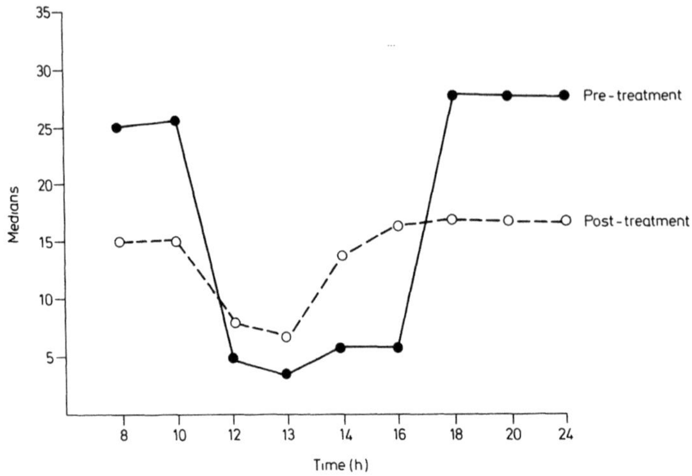

Fig. 2. Results of budipine treatment in a patient with on-off fluctuations during levodopa therapy

The adverse effects of budipine are similar to those of anticholinergics, but they are milder as compared with trihexyphenidyl and benztropine. Two patients who did not tolerate benztropine could tolerate budipine. Among 16 patients, there were 8 with adverse effects, 8 without. All the patients with adverse effects experienced a mild degree of dryness of the mouth. Other side effects included mild blurring of vision, a mild degree of constipation, and slight discomfort during urination in two patients, and orthostatic hypotension in one patient. All these side effects were relatively mild and disappeared soon after reduction of dosage of budipine. Routine analysis of blood and urine was performed according to the therapeutic regimen. No abnormal finding was noted. There was no change of the electrocardiogram before and after treatment with budipine.

Table 1. Therapeutic efficacy of budipine after parenteral application in three parkinsonian patients

Patient No.	Symptom	Dosage	Route	Efficacy	Duration of efficacy of each injection (h)
14	Akinesia and rigidity	20 mg	i.v.	$18 \rightarrow 6^{a}$	24-48
15	Tremor	20 mg	i.v.	Amelioration	24-48
16	Akinesia and rigidity	20 mg	i.m.	$14 \rightarrow 11.5^{a}$	48-72

[a] Webster score.

Discussion

Quite good therapeutic results were obtained with the advent and wide clinical application of levodopa preparations such as Madopar and Sinemet in parkinsonian therapy and they have remained the most effective medication for Parkinson's disease up to the present. However, long-term administration, especially with large doses, may lead to adverse effects manifested by dyskinesias or oscillations in performance. Moreover, the therapeutic effect of levodopa preparations usually regresses after 2-4 years. This is why, according to several authors, one should postpone the initiation of levodopa administration if possible and use the minimal effective dose (Fahn and Calne 1979; see the chapter by Yahr). As a further measure to increase the efficiency of drug therapy for parkinsonism, new medications are evaluated to be used concomitantly with levodopa when oscillations in performance and reduction of therapeutic effect of levodopa set in. Budipine may have some influence on dopamine release and uptake and may also possess anticholinergic properties and exert a stimulatory effect on the serotonergic system. In our series of 16 cases of Parkinson's disease with tremor or akinesia and with on-off phenomena and dyskinesias during the course of treatment with levodopa, we noticed rather good therapeutic effects after treatment with budipine in 15 cases. Budipine has relatively few adverse reactions, mainly dryness of the mouth. These reactions are tolerated by most patients. No psychiatric symptoms were noticed. With budipine, we have not found the side effects associated with bromocriptine and lisuride such as lassitude, dizziness, psychiatric symptoms, and thrombocytopenia (Wang 1982; Luo 1982). That is why we are of the opinion that budipine is a relatively safe drug when used in combination therapy. Further observation and study are needed to see how long the therapeutic effect of budipine lasts.

Summary

Sixteen cases of Parkinson's disease treated with budipine are reported. There were ten cases with rigidity and akinesia and four cases with tremor as the predominant symptoms, and two cases with on-off phenomena and dyskinesia during the course of treatment with levodopa. In the latter two cases, budipine treatment was given for 1.5 years and still has a good effect. The minimal daily dose of budipine was 10 mg p.o. and the maximal 50 mg; the daily dose was 30 mg in the majority of cases. Three patients were treated by i.v. and i.m. injections of budipine and showed good improvement in akinesia and tremor. Efficacy of each injection lasted for 24-48 h. No abnormal change was found in blood and urine routine parameters or electrocardiogram before and after treatment with budipine. Side effects of budipine noticed in this series of patients included a mild degree of dryness of the mouth in eight patients, blurring of vision, slight uncomfortable sensation during urination, and a mild degree of constipation in two patients, and orthostatic hypotension in one patient. Budipine is an effective and relatively safe drug to be recommended in the combination therapy of Parkinson's disease.

References

Fahn S, Calne DB (1978) Consideration in the management of parkinsonism. Neurology (Minneap) 28: 5-8

Menge HG, Brand U (1982) Pharmacological summary of Budipine, a new 4,4-diphenylpiperidine derivative used in treating Parkinson's disease. Arzneimittelforsch 32: 85-98

Luo Y et al. (1982) The drug therapy of primary Parkinson's disease in the elderly. Clin J Geriatr 1: 129

Wang X (1982) The clinical observation of the therapeutic effect of Lisuride in treatment of Parkinson's disease. Clin J Nerv Ment Dis 8: 209

Long-Term Treatment with Budipine

F. Oppel

Introduction

Studies on the effect of budipine in Parkinson's disease were started in 1976 (Brandt and Oppel 1979; Oppel and Umbach 1977). We have investigated budipine's influence on akinesia and tremor in akinesia tests, pulmonary function tests, and by quantitative tremor measurements (Oppel and Umbach 1977). Budipine was administered in combination with other antiparkinsonian drugs. The major result fo these tests was that budipine had an outstanding effect on tremor, whereas the effect on akinesia was not significant.

Patients and Methods

Patient selection was primarily on the basis of tremor. We involved 25 patients with bilateral tremor symptoms, 9 women and 16 men, with an average age of 66 years (55-80 years). All were undergoing combined antiparkinsonian treatment, in the majority L-dopa drugs and anticholinergics, with inadequate success. Budipine was administered in addition to the previous medication, orally, at doses between 40 and 60 mg/day.

We have devised our own technique for quantitative measurement of budipine's antitremor effect (Oppel and Umbach 1977). A piezoelectric element was placed between

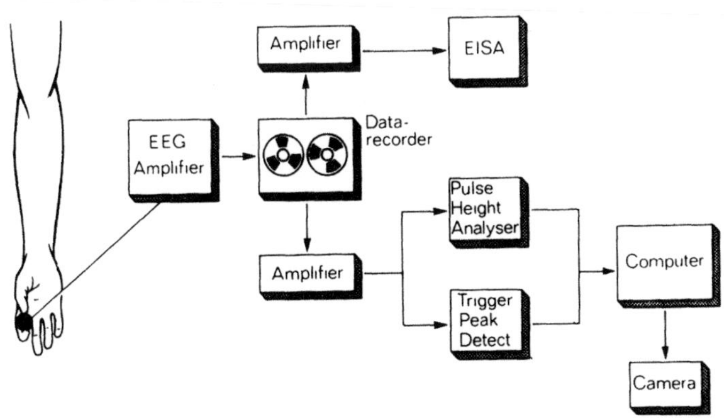

Fig. 1. Apparatus employed in quantitative analysis of tremor frequency and amplitude

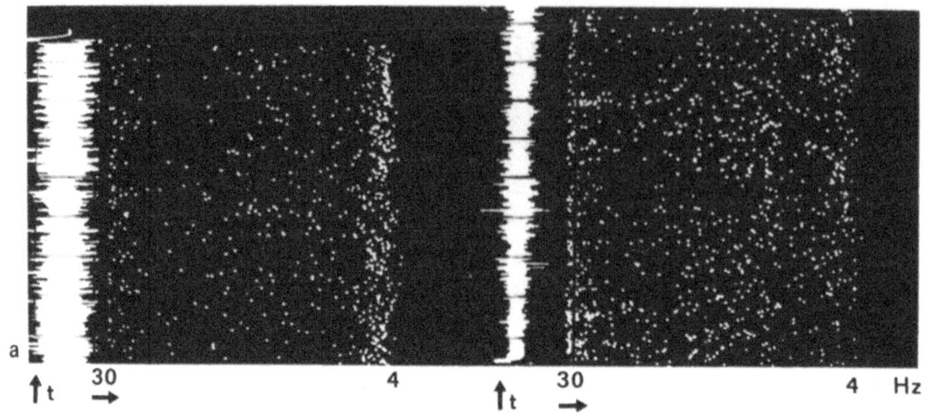

Fig. 2 a, b. Original EISA-grams **a** before; and **b** after budipine treatment. Time (t = 100 s) is plotted on the ordinate and frequency (30-4 Hz) on the abscissa. The ordinate also indicates tremor amplitude. The large frequency band at 4-6 Hz before treatment is replaced by higher frequencies after treatment (see also Fig. 3)

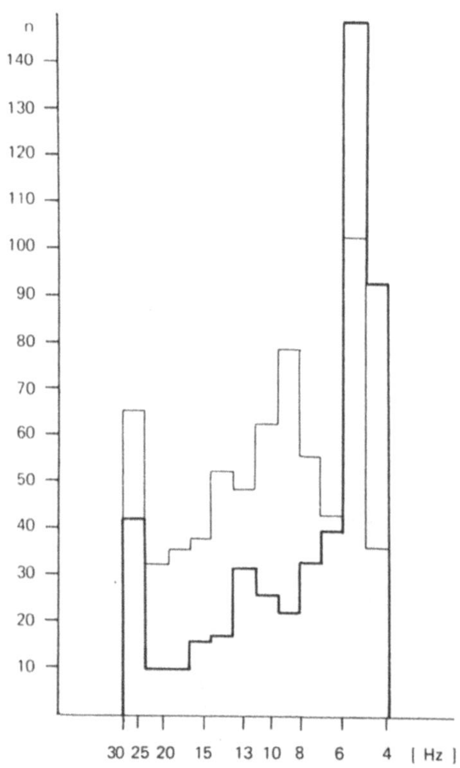

Fig. 3. Frequency histogram of the curves shown in Fig. 2. Before *(thick line)* and after budipine treatment, mean frequency is increased from 9.8 to 12.7 Hz. N = number of cycles per 100 s

the patient's thumb and index finger and fixed with adhesive tape. The tremor curve was recorded via an EEG unit with preamplifier, and stored on tape. The period of recording was 180 s (80 s of habituation and 100 s of analysis). Frequency analysis was done by an EEG interval spectrum analyzer (EISA) (Tönniss 1969) with a lower limit of 0.5 Hz and a deflection of 10 s/cm over a period of 100 s. The amplitude was analyzed by an analog-digital converter (ADC) equipped with a signal averager and intermediate elements, e.g., pulse height analyzer and histogram signal shaper. Triggering was by "peak detect" (amplitude maximum) from the signal shaper. The resolution of the histogram was equivalent to an amplitude chage of 1.7 mV per address (Fig. 1).

Results and Discussion

Frequency Analysis

During frequency analysis, aperiodic cycles were measured, digitalized, reciprocated as frequency over 1-30 Hz, and shown on the oscillograph as corresponding points. Figures 2 and 3 give typical EISA-grams before and after therapy, and frequency histograms. In the 25 patients in whom tremor was recorded on both hands, mean frequency before budipine medication was 9.8 Hz, rising to 12.7 Hz after therapy. The relatively high mean frequency before therapy is explained by the fact that tremor was recorded from the extended arm which provokes an action tremor of higher frequency, resulting in an upward shift of the mean anticipated resting tremor of between 4 and 6 Hz (Fig. 2).

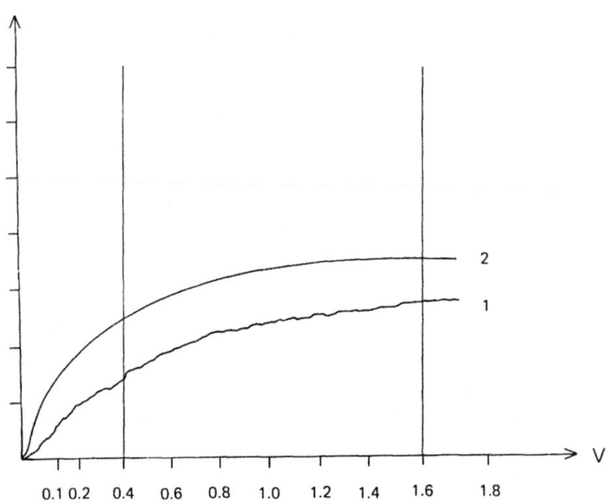

Fig. 4. Cumulative frequency of parkinsonian amplitude before *(curve 1)* and after budipine treatment *(curve 2)* (original recording, see Fig. 5). Typically, the curve after treatment shows a steeper initial rise, consistent with a larger number of low amplitudes. N = number of amplitudes, V = voltage

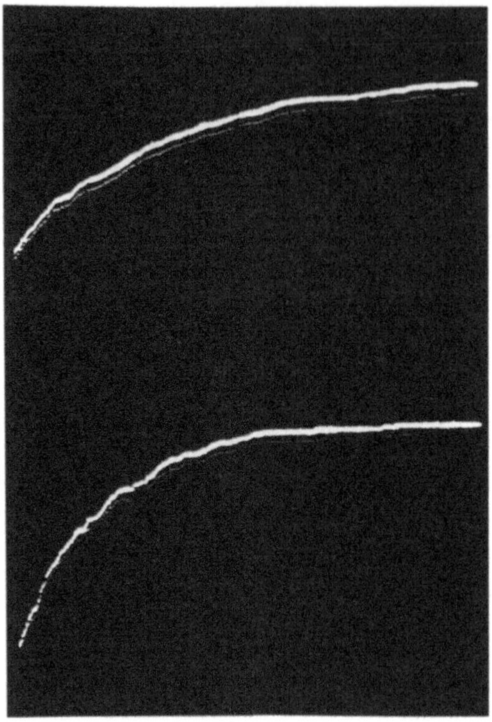

Fig. 5. Original amplitude histogram before *(upper curve)* and after budipine treatment *(lower curve)* (see also Fig. 4)

Amplitude Histograms

Amplitude histograms of physiologic tremor show peak rises in the lower amplitude range, leveling off to nearly horizontal. The more pronounced parkinsonian tremor, the flatter the curve. Thus, the severity of tremor and clinical improvement are easily visible (Figs. 4 and 5). Comparative measurements were done in the lower range at 400 mV and in the higher range at 1600 mV (Fig. 4). The quotient of the two (1600/400) was obtained, providing information on the segment between 400 and 1600 mV. Thus, an approximation to 1 corresponds to a rarefaction of amplitude segments between the two marks. An approximation to 1, together with a relatively high value at 400 mV, is characteristic of physiologic tremor (Fig. 4). The mean value (quotient) for the 25 patients undergoing budipine treatment improved from 2.17 to 1.69, which indicates an approximation to physiologic tremor with marked clinical improvement. Figure 5 gives original tracings, which clearly show the steep rise after budipine administration (larger number of lower amplitudes). Table 1 summarizes the results of both frequency and amplitude analysis.

Another, specific study by this same technique in stereotactic patients (Brandt and Oppel 1979) gave the following result. Defining amplitudes up to 400 mV as low and those between 400 and 1600 mV as high, the portion of low amplitudes for 17 patients before stereotaxy was 38%, immediately afterwards 79%, and 1-9 years later 60%. It was possible with budipine to increase the number of low amplitudes from 55% to 71% (Table 2). We were able to monitor 12 of the 25 patients over a period of 3-6 years, with regular follow-ups. Here are the details of this long-term treatment.

Table 1. Results of frequency and amplitude analysis before and after budipine treatment. The increase in mean tremor frequency by 2.9 Hz and the decrease in the quotient by 0.48, equivalent to amplitude diminuation, indicate approximation physiologic tremor

	Means (both hands) N = 25, N = 25		
	Before treatment	After treatment	Difference
Frequency (Hz)	9.8	12.7	+2.9
Amplitude (quotient)	2.17	1.69	- 0.48

Table 2. Results after stereotactic subthalamotomy in terms of lower tremor amplitudes. Following a relapse of tremor in some patients, wc were able to raise the proportion of low amplitudes again with budipine in five patients

Stereotaxy N = 17	Low amplitudes (up to 400 mV)	Budipine treatment N = 5	Low amplitudes (up to 400 mV)
Before	38%	Before	55%
Immediately after	79%	After	71%
After 1-9 years	60%		

Dosage

Three patients were given 40 mg budipine throughout, two patients took 60 mg, and one patient 90 mg/day. In five patients, the initial dose of 40 mg was increased to 60 mg and in one further patient to 80 mg.

Concomitant Medication

In all cases, budipine was administered in combination with other antiparkinsonian agents, viz., in six patients together with anticholinergics, L-dopa + benserazide and amantadine, in two patients with anticholinergics and L-dopa plus benserazide, in one patient with anticholinergics and amantadine, in two patients with anticholinergics alone, and in one patient with L-dopa plus benserazide alone.

Concomitant Disease

All patients presented with heart failure, five suffered from a mild to moderate hypertension, and one from diabetes mellitus. Three patients were depressed, one was occasionally confused.

Long-Term Effect

There was a uniformly marked effect on tremor in all patients, in two cases setting in only after dosage had been increased to 60 mg, and to 80 mg in one other case. Four patients reported a continuing marked improvement in mood, mostly in the sense of greater psychic stability, two of these had been depressed before going on budipine.

Tolerance

On the whole, budipine tolerance up to 6 years was extremely good. Side effects were one case of inappetence at 90 mg/day, which disappeared after dose reduction to 60 mg. A metallic taste reported in another case under 60 mg/day proved fully reversible.

Incompatibilities with other antiparkinsonian drugs were not reported, nor was there aggravation of known side effects following other concomitant drugs. There was no influence on concomitant diseases, no change in biochemical parameters of ESR, blood counts, blood glucose, GOT, GPT, alkaline phosphatase, and creatinine.

Summary

Budipine appears to be a suitable treatment for Parkinson's disease. Its effects at doses of 40-90 mg/day are evident primarily on tremor, including recurrent tremor after stereotactic subthalamotomy, and mood. In some cases, patients experienced a pleasurable sense of stability. Restlessness or confusion were not observed. With administration over a period of up to 6 years, efficacy did not decline, regardless of age, duration of illness, previous history, or concomitant diseases regularly observed with old age, such as heart failure, hypertension, or diabetes mellitus. We have administered budipine without exception in combination with other antiparkinsonian drugs and, despite the large doses of these druggs in individual cases, there was an additional effect on tremor. Tolerance was good. Except for the two reversible cases, no side effects were observed on oral administration of 40-90 mg/day, even over prolonged periods.

References

Brandt F, Oppel F (1979) Quantitative measurement of parkinsonian tremor before and after stereotactic operation. Neurosurg 7: 197-202

Oppel F, Umbach WU (1977) A quantitative measurement of tremor. Electroencephalogr Clin Neurophysiol 43: 885-888

Tönnies JF (1969) Automatische EEG-Intervall-Spektrumanalyse (EISA) zur Langzeitdarstellung der Schlafperiodik und Narkose. Arch Psychiatr Nervenkr 212: 423-445

Budipine Treatment of Neuroleptic-Induced Parkinsonism

S. Bien and R. Wörz

Introduction

Extrapyramidal disturbances of movement are among the most frequent undesirable concomitant effects of neuroleptic therapy (Ayd 1961; Benkert and Hippius 1980; Haase 1982). It is indisputable that early dyskinesia can be favorably influenced by anticholinergics (e.g., biperiden or methixene). The neuroleptic-induced parkinsonian syndrome, on the other hand, presents a difficult problem. The discussion is still not concluded as to whether or not the combined administration of neuroleptics and anticholinergics raises the risk of developing late dyskinesia (Haase 1982 and others). This point is particularly important because there is presently no dependable method for treating late dyskinesia (Neumann 1982). The development of drugs with better efficacy than that of anticholinergics or with an alternative approach is thus of substantial clinical importance, especially for chronic schizophrenics under long-term neuroleptic treatment.

Patients and Methods

Included in this study were 30 patients with a drug-induced parkinsonian syndrome. There were 8 male and 22 female patients aged 29-79 years (median 60.5 years). Figure 1 shows the age distribution. Of the 30 patients, 27 were undergoing treatment for chronic schizophrenia in the hospital. Long-term neuroleptic treatment was necessary for two cases of severe psychopathy and one case of pronounced mental deficiency. Table 1 shows the distribution of diagnoses and the duration of each disease.

The patients were treated with budipine from 4 weeks to a maximum of 3 years (median 12 months). To begin with, they received 3x10 mg budipine orally. The medication was increased to 3x20 mg in most cases on the 4th day; in individual cases, the dose was increased to 3x20 mg on the 8th day.

The treatment results were initially documented at intervals of 1 week; after 3 weeks they were recorded every month. The neurologic findings were documented in each case according to the Webster rating scale. Efficacy and tolerance were assessed by both the investigator and patients, based on the scale: very good, good, satisfactory, moderate, and poor.

The following laboratory parameters were determined prior to treatment, then at weekly intervals in the beginning, and finally at intervals of 4 weeks after the first month: hemoglobin, leukocytes, SGOT, SGPT, LDH, alkaline phosphatase, cholesterol,

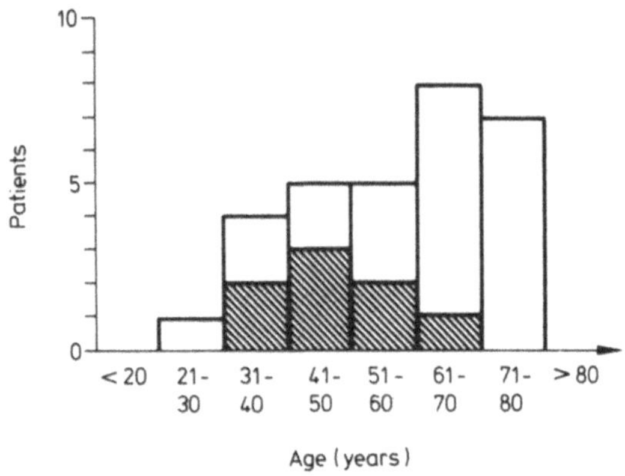

Fig. 1. Age distribuction of 30 patients *(hatched area* represents males)

Table 1. Diagnoses and duration of disease (N = 30)

Diagnosis	Duration of disease (years)	Number of patients
Chronic schizophrenia	< 5	1
	5-10	5
	> 10	13
Catatonic schizophrenia	> 10	1
Chronic productive schizophrenia	> 10	2
Paranoid hallucinative schizophrenia	< 5	2
Pfropfschizophrenia	5-10	1
Psychopathy	> 10	2
Somatogenic depression	> 10	1
Endogenic depression during involutional age	5-10	1
Severe mental deficiency	> 10	1

glucose, urea, erythrocyte sedimentation rate, potassium, calcium, neutral fats, and urinalysis. An ECG was taken with standard and precordial leads every 4 weeks.

Results

Not all the symptoms on the Webster rating scale were pathologic in each patient prior to treatment with budipine. Table 2 contains the number of patients with the symptoms described that were pathologic before treatment with budipine. The extent of change in the individual parameters is not contained in this survey. The assessments of the 1st, 2nd, and 3rd week are shown in Figs. 2, 3 and 4.

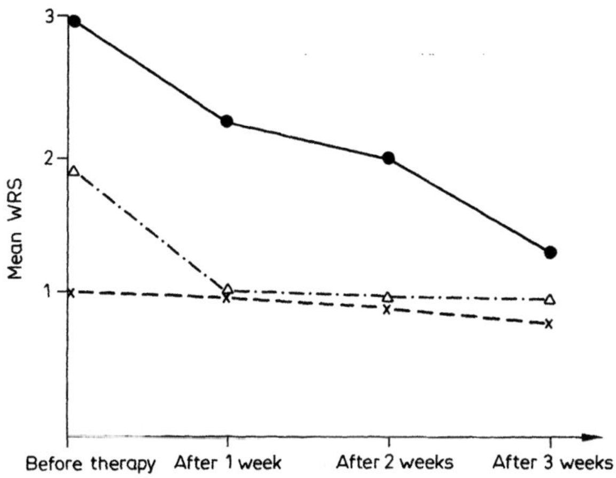

Fig. 2. Changes in arm-swing *(full line* $P < 0.001$), gait *(broken line* $P < 0.01$), and tremor *(dashed-dotted line* $P < 0.001$), under budipine treatment, according to Webster rating scale

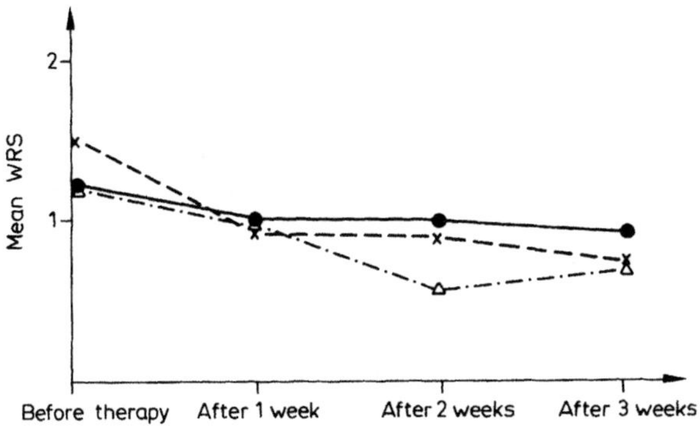

Fig. 3. Changes in bradykinesia *(full line* $P < 0.001$), rigidity *(broken line* $P < 0.001$), and posture *(dashed-dotted line* $P < 0.001$) under budipine treatment, according to Webster rating scale

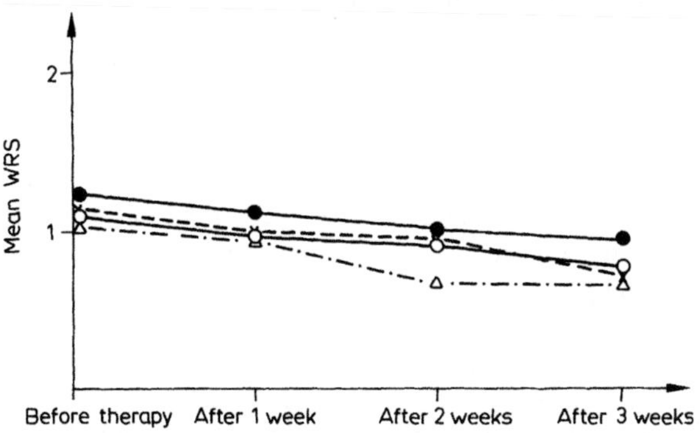

Fig. 4. Changes in hypomimia *(full line, full circles* $P < 0.01$), seborrhea *(broken line* $P < 0.001$), voice change *(dashed-dotted line* $P < 0.01$), and self-care *(full line, open circles* $P < 0.05$) under budipine treatment, according to Webster rating scale

Table 2. Incidence of individual symptoms (Webster rating scale)

Symptom	No. of patients	Symptom	No. of patients
Tremor	24	Seborrhea	20
Hypomimia	24	Insufficient arm-swing	19
Bradykinesia	23	Parkinsonian gait	19
Rigidity	22	Voice change	15
Posture	22	Limited self-care	15

Tremor, seborrhea, bradykinesia, rigidity, posture, and arm-swing improved highly significantly after 3 weeks ($P < 0.001$). Facial change up to amimia, parkinsonian change in speech, and parkinsonian gait also improved significantly after 3 weeks ($P < 0.01$), as did self-care ($P < 0.05$). No decline in drug effect was observed during further treatment of the patients.

The following side effects were observed during treatment with budipine: two patients complained of slight restlessness; another two experienced slightly pronounced fatigue. The main group of deteriorations occurred as psychotic exacerbation in patients with chronic or partly chronic productive schizophrenia. This was observed in six patients during the entire study (including the long-term study). Table 3 shows the side effects with frequency distribution of severity and with the course of the side effect. In addition, there were complaints of dizziness (four patients), restlessness (two patients), and fatigue (two patients). Slight dryness of mouth was reported three times. Table 3 also shows that most of the side effects mentioned subsided while continuing treatment with the antiparkinsonian drug. The treatment was discontinued in six patients who had initially been stabilized on budipine. The reasons given for two of these patients were physical diseases, independent of the treatment with budipine. Treatment was stopped on days 24 and 60 for ethical reasons. In four other patients, the treatment was stopped because of psychotic exacerbation. All these patients were chronic productive schizophrenics who had described recurrent psychotic exacerbations in their history. A connection with budipine treatment was thus possible, but it could not be proven; an activation of the basic disease process could not be excluded. Here, too, budipine treatment was discontinued for ethical reasons.

In the synopsis and in the statistical analysis of the observed laboratory parameters there was no significant change in the values, not even during long-term treatment. Slight deviations or signs of other diseases, such as occasional intercurrent infections or urinary tract infections, turned out to be independent of budipine treatment. ECG anomalies did not occur during budipine treatment; preexisting ECG changes did not show any tendency toward dterioration.

Both investigator and patients gave a summarized assessment of the efficacy and tolerance of the drug. The results are given in Tables 4 and 5. The investigator described the effect as being good to very good in 60% of the patients. The patients found the effect to be good or very good 66.6% of the time. The investigator observed one case of poor efficacy; in no case did the patients report poor efficacy. The investigator designates

Table 3. Side effects and their course during budipine therapy[a]

Side effect	Degree				Connection with budipine			Course
	Slight	Moderate	Severe	Total	Improbable	Questionable	Probable	
Restlessness	2			2	1	1		1
Fatigue		2		2		2		1
Deterioration of psychic symptoms	4	2		6		3	3	1x1 2x2 1x3 2 N.sp.
Dizziness	4			4		4		1
Dryness of mouth	3			3		2	1	2x3 1x2

[a] *1* Symptoms subside spontaneously under furhter treatment.
2 Symptoms subside under specific treatment.
3 Symptoms are tolerated.
4 Symptoms are tolerated after reducing dose.
5 Symptoms subside completely after reducing dose.
6 Treatment had to be stopped.
N.sp., not specified.

Table 4. Investigator's and patients' concluding assessment of budipine's efficacy

Assessment of efficacy	Total No. (%)	Very good (%)	Good (%)	Satisfactory (%)	Moderate (%)	Poor (%)	No information (%)
Investigator	30	6	12	5	6	1	0
	(100)	(20)	(40)	(16.7)	(20)	(3.3)	(0)
Patients	30	10	10	5	2	0	3
	(100)	(33.3)	(33.3)	(16.7)	(6.7)	(0)	(10)

Table 5. Investigator's and patients' assessment of tolerance

Assessment of tolerance	Total No. (%)	Very good (%)	Good (%)	Satisfactory (%)	Moderate (%)	Poor (%)	No information (%)
Investigator	30	16	11	0	1	2	0
	(100)	(53.3)	(36.7)	(0)	(3.3)	(6.7)	(0)
Patients	30	17	10	0	0	0	3
	(100)	(56.7)	(33.3)	(0)	(0)	(0)	(10)

the tolerance as being good to very good in 90% of the patients, moderate in 3.3%, and poor in 6.7%. The patients found the tolerance to be subjectively very good in 56.7% of the cases and good in 33.3%. None of the patients described moderate or poor tolerance. No information on tolerance was available in three cases.

Discussion

Budipine (3x20 mg/day) proved to have a good to very good effect on the great majority of patients with neuroleptic-induced parkinsonian syndrome. An improvement in parkinsonian symptoms was already noted during the closely monitored 3-week initial period of the study; using the Webster rating scale, a clear and statistically highly significant regression of parkinsonism was obtained. An improvement in self-care was experienced by a large number of the patients.

The slightly stimulating quality of budipine appears to be relevant to the exacerbation of chronic schizophrenic psychoses. Budipine should thus not be used in patients with productive psychotic symptoms and, above all, with hallucinations. The stimulating effect, which increased drive to the point of slight restlessness in two patients, is not necessarily a disadvantage in schizophrenic patients with lack of drive, affective impairment, indifference, and social withdrawal. Three patients complained of slight dryness of the mouth. Pronounced hypersalivation existing beforehand in one female patient subsided markedly; the side effect in this case was thus not undesirable.

Conclusions

According to our observations, budipine is a welcome alternative for treating neuroleptic-induced parkinsonian syndromes. Because of the described haloperidol antagonism (Menge and Brand 1982), care should be exercised with productive psychotic patients. Our experience shows that psychotic exacerbation cannot be excluded. A relatively specific side effect is dryness of the mouth or reduced salivation.

References

Ayd FJ (1961) A survey of drug-induced extrapyramidal reactions. JAMA 175: 1054-1060

Benker O, Hippius H (1980) Psychiatrische Pharmakotherapie, 3rd edn. Springer, Berlin Heidelberg New York

Haase H-J (1982) Therapie mit Psychopharmaka und anderen, das seelische Befinden beeinflussenden Medikamenten. 5th edn. Schattauer, Stuttgart New York

Menge HG, Brand U (1982) Pharmacological summary of budipine, a new 4,4-diphenylpiperidine derivative used in treating Parkinson's disease. Arzneimittelforsch 32/1: 2, 85-98

Neumann N-U (1982) Behandlung der Spätdyskinesien. In: Lungershausen, E, Wörz R (eds) Interdisziplinäre Problemgebiete. Referate der Jahresversammlung der Bayr. Nervenärzte 1979. Günzburger Schriften zur klinischen Psychiatrie, vol 4, pp 88-100

Randomized Double-Blind Trial of Budipine and Biperiden in Patients with Neuroleptic Extrapyramidal Symptoms

P. König and I. Pass-Kosmath

Introduction

Preclinical studies proved the piperidine-derivative, budipine to exert strongly antagonistic properties against tremorine and oxotremorine (Menge and Brand 1982). These effects were present at low dosages and cannot be explained by the mild anticholinergic potency of the compound. In addition, budipine counteracts the cataleptic effects of monoamine depletors, such as reserpine; also dopamine and norepinephrine antagonism of haloperidol or phenothiazines is reduced. Budipine does not compete with spiroperidol at the binding site (Menge and Brand 1982). Animal experiments suggest that the drug stimulates catecholaminergic neurons, leading to higher vigilance and motility of the treated animal. However, the usual autonomous side effects in the action of psychostimulants were not detected in these experiments. Budipine thus being identified as a possible antiparkinsonian agent, it seemed feasible to test the compound in patients with Parkinson's disease and also to test its efficacy on extrapyramidal side effects (EPS) induced by neuroleptic treatment.

Methods

A randomized double-blind trial was scheduled for 30 inpatients at a psychiatric hospital.The group comprised members of both sexes who gave their written consent. Long-term neuroleptic pretreatment was considered to ensure manifest EPS and possibly forms of tardive dyskinesia, though this disorder is disregarded in this part of the study. Psychiatric diagnosis comprised the International Classification of Diseases categories of schizophrenia (ICD 295.0-295.9). Neuroleptic medication was continued, irrespective of the type of drug, as was all other essential medication, except drugs that might influence vigilance or motor performance (e.g. tranquilizers). Antiparkinsonian agents were washed out in a 3-day washout phase (Tables 1 and 2).

Extrapyramidal syndromes were present in all patients; in one for over 1 year, in five up to 5 years, in three up to 10 years, and in ten patients for over 10 years. For exclusion criteria, we decided on severe pulmonary, cardiac, renal, gastrointestinal or liver disease; severe cerebrovascular impairment; (neurologic) movement disorders, epilepsy, pregnancy, or breast feeding. For ethical reasons, we decided against a placebo-controlled study. Thus after a 3-day washout, one of the drugs, budipine or biperiden (we are indebted to Byk Gulden, Konstanz, Federal Republic of Germany, for identical preparation and generous supplies of both compounds), was given double-blind,

Table 1. Demographic data of both groups of patients

	Budipine	Biperiden
Number of patients	10	8
Mean age (years)	43 (20-73)	43 (26-67)
Male	4	4
Female	6	4
Body weight (Broca's index)		
Normal (80%-110%)	5	4
Overweight (> 110%)	5	3
Underweight (< 80%)	0	1

Table 2. General condition, duration of psychiatric illness (schizophrenia), and type of continued neuroleptic medication

	Budipine	Biperiden
General condition		
Good	8	8
Medium	1	
Poor	1	
Psychiatric illness (ICD 295.0-295.9)		
1- 5 years	4	1
6-10 years	1	2
> 10 years	5	5
Concomitant neuroleptics		
Phenothiazines	6	6
Other tricyclic neuroleptics	3	2
Other oral neuroleptics	5	5
Depot neuroleptics	2	
Neuroleptic dose regimen		
Stable	8	12
Decrease	2	
Increase	1	
Stopped	1	
Begun	2	1
Depot injections	2	

first in a fixed-dose regimen for 1 week, later in a flexible-dose regimen for the following 7 weeks, the duration of the study being 8 weeks in all (Table 3).

The following assessments and scales were used for baseline and after 1, 4, 6 and 8 weeks: Webster rating scale (Webster 1968), Gerlach's (1979) and Simpson's (1979) tardive dyskinesia scales, Grünberger (1977) motor activity test, drawing and writing samples, and flicker fusion frequency analyses. A self-rating scale for various (motor) activities and an overall individual and clinical assessment of drug efficacy was applied. In addition, clinical and laboratory data were gathered.

Table 3. Dosage regimen in mean daily doses of budipine and biperiden

	Mean daily dose (mg)	
	Budipine	Biperiden
Week 1	66.0 (40-120)	6.3 (2-10)
Weeks 2-3	70.0 (40-120)	6.8 (2-10)
Weeks 4-6	74.0 (40-120)	7.0 (2-10)
Weeks 7-8	72.0 (40-120)	7.0 (2-10)

Results

At the present date, the Webster scale, self-rating, and flicker fusion results of 18 probands have been calculated and are presented. The initial Webster scale data (Tables 4 and 5) show a slight bias toward severe forms of EPS in the budipine-treated group. Improvement is documented toward the end of the trial period. This shows up more clearly in the Webster scores decreasing by 0.6, for budipine, as opposed to 0.3 for biperiden. Also in the self-assessment scale, improvement of impairment is documented as more pronounced in the second phase of the trial, as regards budipine treatment (Table 6). Untoward side effects such as dizziness, fatigue, sleeplessness, nervousness, and mild agitation were only rarely observed and practically evenly distributed in both treatment groups (Table 7).

Mean values of flicker fusion frequency are shown in Table 8. Wilcoxon's test showed a significant difference in the results of budipine-treated patients, as compared with

Table 4. Severity of EPS according to Webster rating at the beginning and end of study

	Budipine	Biperiden
Effects of anticholinergic pretreatment of EPS		
Good	8	5
Medium	2	1
No data		2
Severity of EPS, Webster baseline		
Absent (0)	0	1
Minimal (0.1-1.0)	8	7
Medium (1.1-2.0)	2	
Severe (2.1-3.0)		
Severity of EPS, Webster end of trial		
Absent	2	4
Minimal	7	4
Medium	1	
Severe		

Table 5. Improvement of EPS symptoms according to the Webster scale

Symptoms	Budipine	Biperiden
Bradykinesia	2	2
Rigidity	3	2
Tremor	2	1
Posture	4	0
Concomitant movement	1	1
Walk/gait	1	0
Speech	1	0
Seborrhea	1	3
Facies	3	1
Self-reliance	1	0
Total	19	10

Table 6. Self-assessment of improvement in condition and EPS-related impairments and symptoms

Self-rating	Budipine			Biperiden		
	Week 1	Week 4	Week 8	Week 1	Week 4	Week 8
General condition	5	2	4	0	0	1
Eating (motor performance)	1	1	0	2	1	0
Drinking (motor performance)	0	0	0	1	1	0
Writing	0	1	1	0	0	1
Walking	0	0	0	0	0	0
Dressing/personal hygiene	0	0	0	0	0	0
Tremor	0	0	3	0	1	0
Rigidity	0	1	1	1	1	1
Speech	0	0	0	1	0	0
Perspiration	0	1	1	0	2	1
Mood	2	1	3	2	2	0
Total	8	7	13	7	8	4

Table 7. Untoward effects of the different compounds as rated by the patients. Budipine shows a relatively higher frequency of side effects at later stages of therapy (yes/no answers)

Symptoms	Budipine			Biperiden		
	Week 1	Week 4	Week 8	Week 1	Week 4	Week 8
Nausea	0/10	0/10	1/ 9	1/ 7	0/ 8	0/ 8
Vomiting	0/10	0/10	1/ 9	0/ 8	0/ 8	0/ 8
Vertigo	2/ 8	3/ 7	3/ 7	0/ 8	0/ 8	0/ 8
Fatigue	1/ 9	4/ 6	3/ 7	1/ 7	2/ 6	1/ 7
Insomnia	1/ 9	1/ 9	0/10	0/ 8	0/ 8	0/ 8
Nervousness	2/ 8	3/ 7	1/ 9	2/ 6	2/ 6	2/ 6
Total	6/54	11/49	9/51	4/44	4/44	3/44
Relative frequency of side effects	0.11	0.22	0.18	0.09	0.09	0.07

Table 8. Values of flicker fusion frequency for the different treatment groups (Wilcoxon-Pratt test)

Budipine						
n	9	9	9	9	9	9
Median	43.10	43.20	42.75	41.30	43.35	43.35
Minimum	41.15	38.35	37.90	38.95	38.20	38.20
Maximum	48.05	44.10	44.75	43.55	47.00	47.00
	Ref.	+	+	+	+	+
Biperiden						
n	8	8	8	8	8	8
Median	43.60	43.07	42.88	42.30	42.88	44.28
Minimum	41.85	39.65	41.20	40.25	40.65	39.50
Maximum	48.70	45.90	46.95	44.25	48.05	48.05
	Ref.	N.S.	+	+	N.S.	N.S.

Ref., reference or baseline; N.S., not significant

biperiden. Budipine treatment leads to a statistically significant rise of the flicker fusion parameters. Clinical and laboratory data such as liver enzyme function, electrolytes, haematologic counts, and ECG were all within (age-dependent) normal limits.

Discussion

In the treatment of neuroleptically induced EPS, biperiden together with procyclidine (and to a lesser extent bornaprin) is one of the most widely applied drugs. Therefore, our choice of biperiden as collateral for a double-blind trial against budipine seems justified, the efficacy of the drug in this indication is well documented (e.g., Goodman and Gilman 1980). After conducting a randomized double-blind study on inpatients of both sexes of budipine against biperiden, we consider budipine to be practically equally effective.

Our budipine-treated group starts with a slighlty inferior Webster scale rating thus being more disabled as compared with the biperiden-treated patients. During the first 2 weeks of treatment, both compounds are nearly equal in performance. This alone would suggest slightly better efficacy of budipine. After the third week, the Webster score for budipine-treated patients fell by factor 0.6, for biperiden this decrease is not so apparent (0.3).

Self-assessment of improvement of EPS by budipine treatment shows a more striking difference in evaluation: this substance is considered more effective than biperiden by our probands. Global assessment by doctor's rating also shows a preference for budipine as an antiparkinsonian agent. Global rating of change of symptoms according to the Webster scale before and after treatment with the two drugs gives similar results.

Untoward side effects seem to be evenly distributed in both drugs, with a slight bias towards budipine. In this group, we had one patient complaining of vomiting and nausea, whereas in the biperiden group, only slight nausea was complained of once. (These effects cannot be characterized as drug specific, they may also have been produced by the neuroleptic medication.)

We consider one of the most interesting findings to be the significant gain in flicker fusion frequency under budipine. This may be in accordance with the animal data relating to vigilance, already cited. Flicker fusion frequency gain may be interpreted as an improvement in vigilance. Abnormal results can be related to organic brain dysfunction, as in dementia, or functional disruptions, as in drug-induced states of reduced consciousness. Another type of reduction of flicker fusion frequency is found in depressive states related to functionally impaired tool performance (Berner 1977). Thus, a compound capable of augmenting flicker fusion performance, may exert certain antidepressive properties, which ought to be further investigated. Another interesting feature in this context are the, albeit minimal, anticholinergic properties of the substance, as some investigators cannot discount acetylcholinergic mechanisms in the etiology of depression (Beckmann 1978).

References

Beckmann H (1978) Biochemische Grundlagen der endogenen Depression. Nervenarzt 49:557-568

Berner P (1977) Psychiatrische Systematik, 1st edn. Huber, Bern Stuttgart Vienna, pp 23, 203

Gerlach J (1979) Tardive dyskinesia. Dan Med Bull 26:5

Goodman LS, Gilman AG (1980) The pharmacological basis of therapeutics, 6th edn. Macmillan, New York, pp 484-486

Grünberger J (1977) Motorischer Aktivitätstest. In: Grünberger J (ed) Psychodiagnostik des Alkoholkranken. Maudrich, Vienna

Menge HG, Brand U (1982) Pharmacological summery of Budipine. Drug Res 32:2, 85-98

Simpson GM (1979) A rating scale for tardive dyskinesia. Psychopharmacology 64: 171-179

Webster DD (1968) Critical analysis of the disability in Parkinson's disease. Mod Treatment 5: 257-282

The Effect of Budipine on the Rheology of the Blood

E. Ott and H. Lechner

Introduction

The main goal of treatment of Parkinson's disease is substitution of neurotransmitter deficit. However, recently hemodynamic factors and the influence of vascular risk factors have also been discussed. The treatment of parkinsonism with L-dopa has been shown not only to improve brain metabolism, but also to increase cerebral blood flow, however, no influence of L-dopa on the plasma or corpuscular blood constituents has been detected (Fazekas et al. 1983; Lavy et al. 1979). The present work reports the influence of a new antiparkinsonian compound, budipine on the rheological properties of the blood in patients with Parkinson's disease.

Patients and Methods

There were 25 patients with Parkinson's disease (18 females, 7 males) with a mean age of 61 years. Patients were on L-dopa treatment (mean dosage 300 mg/day) and received additional peroral budipine (o.46-0.92 mg/kg). Before budipine treatment was started, in all patients, determination of whole blood viscosity and its parameters were performed, including hematocrit, spontaneous platelet aggregation (PAG), ADP- and adrenaline-induced PAG (ADP-PAG, EN-PAG), and fibrinogen. PAG was determined by photoelectric method and blood viscosity was measured by a cone-plate viscometer at 37°C and at four different shear rates (212, 42, 22, 11 s^{-1}). Fibrinogen was estimated by a standard method and hematocrits were determined by the microcentrifuge technique.

Results

As may be seen from Table 1, there was a slight, but not significant, fall of blood viscosity at the end of the treatment period, amounting to 8%. Since the reproducibility of the measuring system is 6%, the real fall of blood viscosity was 2% while hematocrit increased by 0.02.

Table 2 displays the effect of budipine on spontaneous PAG. The indicators used are the transmission time, i.e., the time at which aggregation occurs in the aggregometer, and the maximal amplitude, i.e., the maximal deviation of the aggregation curve in the

Table 1. Influence of 0.5-0.9 mg/kg budipine on blood viscosity (n = 25) in mPa/s

	Shear rate (s^{-1})			
	212	42	22	11
Before	4.8 ± 0.4	5.9 ± 0.6	7.7 ± 0.8	8.7 ± 1.3
After	4.6 ± 0.3	5.7 ± 0.4	7.4 ± 0.9	8.0 ± 1.7

Duration of treatment 5 weeks, change in hematocrit 0.02.

Table 2. Influence of 0.5-0.9 mg/kg budipine on spontaneous platelet aggregation (n = 15)

	Spontaneous PAG	
	Transmission time (min)	Maximal amplitude (cm)
Before	2.8 ± 0.9	8.4 ± 2.7
After	6.5 ± 0.4*	3.2 ± 0.8*

* Significant differences; duration of treatment 6 weeks.

Table 3. Frequency of irreversible induced platelet aggregation before and after budipine administration (0.5-0.9 mg/kg) (n = 15)

	ADP-PAG (0.3×10^{-9} m/ml)	EN-PAG (0.3×10^{-9} m/ml)
Before	9/15	12/15
After	9/15	9/15

Table 4. Influence of 0.5-0.9 mg/kg budipine on fibrinogen (n = 15)

	Fibrinogen (mg per 100 ml)
Before	378 ± 18
After	356 ± 24

Duration of treatment 6 weeks.

10-min recording period. There was a significant prolongation of transmission time, indicating late occurrence of spontaneous PAG and a significant reduction of maximal amplitude, indicating suppression of PAG.

Table 3 shows that the frequency of ADP-PAG and EN-PAG was not influenced by budipine. As can be seen from Table 4, there was a slight, but not significant fall of fibrinogen.

Discussion

The results presented indicate suppression of spontaneous PAG in patients with Parkinson's disease who were given treatment with budipine; this is accompanied by a slight decrease of the fibrinogen level. There was also a slight decrease of whole blood viscosity, despite an almost unchanged hematocrit. As has been shown previously, evidence of enhanced spontaneous PAG may be found with increasing age. Moreover, the aggregation of platelets as well as of red cells is strongly influenced by fibrinogen and may by exaggerated in the face of high fibrinogen levels and vice versa.

It is striking from this investigation that the suppression of spontaneous PAG was accompanied by a fall of fibrinogen, but that, on the other hand, there was no change in the sensitivity of platelets to ADP or adrenaline following 6 weeks treatment. Thus, it may be concluded that the suppression of spontaneous PAG was due rather to decreasing fibrinogen levels than to budipine itself which seems to be metabolized by the liver. However, a definite conclusion is not possible yet.

References

Fazekas F, Ott E, Marguc K et al. (1983) CBF measurements in parkinson disease. International Symposium on measurement of cerebral blood flow and cerebral metabolism in man, 29 Sept-1 Oct 1983, Heidelberg

Lavy S, Melamed E, Cooper G et al. (1979) Regional cerebral blood flow in patients with Parkinson disease. Arch Neurol 36: 344-348

Lechner H, Ott E (1983) Bedeutung der Hämorheologie in Diagnose und Therapie cerebrovaskulärer Erkrankungen. In: Schnaberth G, Pateisky K (eds) Fortschritte der klinischen Neurologie. Thieme, Stuttgart, pp 1-5

Lechner H, Ott E, Fazekas F (1984) Parkinson with high haemodynamic risk. First Symposium of the European Neuroendocrine Association. 4-7 March, Basle

Ott E, Lechner H (1982) Hemorheologic and haemodynamic aspects of cerebrovascular disease. Path Biol 30: 611-614

Ott E, Lechner H, Aranibar A (1974) High blood viscosity syndrome in cerebral infarction. Stroke 5: 330-333

Ott E, Fazekas F, Tschinkel M et al. (1983) Rheological aspects of cerebrovasular disease. Eur Neurol 22 (suppl 1) 35-37

EMG and Computer Analysis of Patients with Tremor and Rigidity Under Budipine Therapy

S. Krafczyk and H. Hökendorf

Introduction

The quantification of tremor amplitudes is still a prerequisite for the evaluation of the efficiency of different drugs in the treatment of tremor. In this report, we discuss the method and the results of tremor measurement with the aid of electromyographic (EMG) examination which has been further analyzed with a digital computer. The first part of our presentation explains the data analysis procedure; the second part deals with the clinical results.

Apparatus and Standardization

The block diagram of Fig. 1 shows our experimental setup. The EMG signal detected by needle electrodes was amplified by commercially available EMG equipment, supported by an oscilloscope, a pen recorder, magnetic tape, and a digital computer. In our investigation, the analogue signal of the complete EMG examination was stored on magnetic tape. To calculate the individual data of each patient, the computer was off line, connected to the magnetic tape via the amplifier.

Selected parts (curve cycles of 3 s), representing the maximum activity the EMG recording, are fed into the computer (Figs. 2 and 3). Six cycles of each recording were analyzed. The aim was to calculate the mean activity and the standard of each cycle. Therefore, the analogue signal to the computer was continuously digitalized ($f\Delta t$ = 2kHz) by means of two analogue-digital converters and then rectified.

The parameter which best represents the course of tremor is the integral value. Figure 4 depicts the integration of the rectified signal, the amplitudes, being summed. The indices T1, T2, and T3 mark the beginning of each tremor group. There is an increase in slope at these positions. The last value in Fig. 4 is the sum of all amplitudes.

Since six cycles of each EMG recording are analyzed, it is necessary to calculate mean values and standard deviations. EMG 1 (before administration of budipine) was taken as a reference value of 100% and the subsequent deviations of recordings are standardized correspondingly. In order to distinguish between the therapeutic effect and the daily fluctuations as well as the fluctuations during measurement, we had to determine a confidence range, covering the spontaneous variations of tremor. This has been performed with prolonged recording in several patients before therapy.

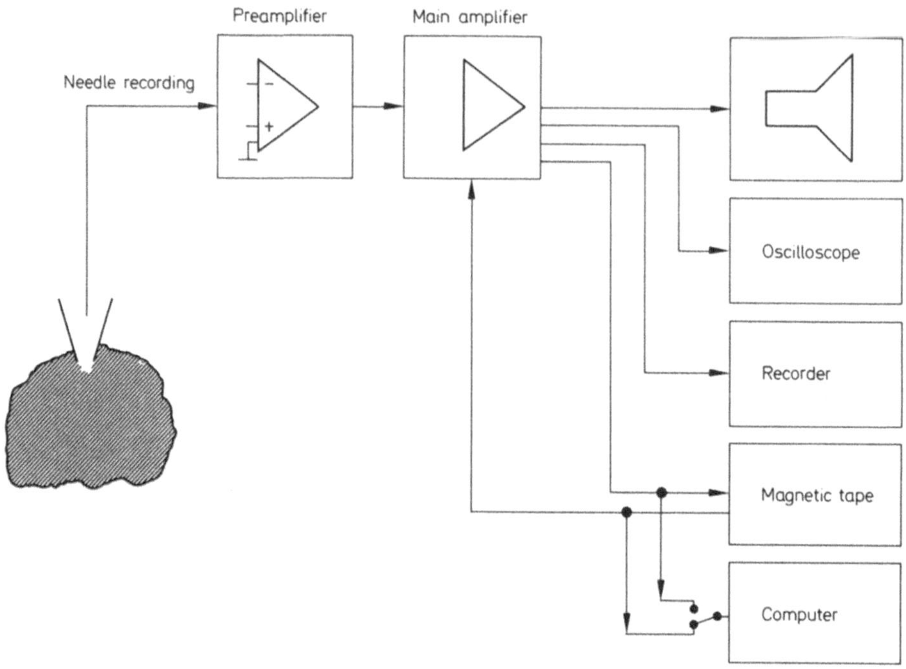

Fig. 1. Computerized EMG apparatus

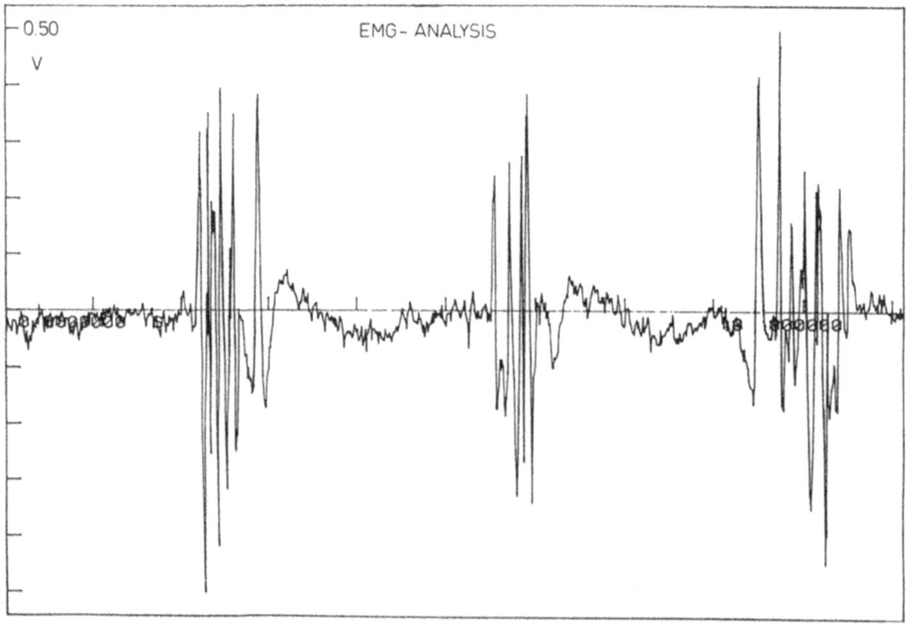

Fig. 2. Muscle action potentials of three tremor groups (abscissa t = 0.4 s)

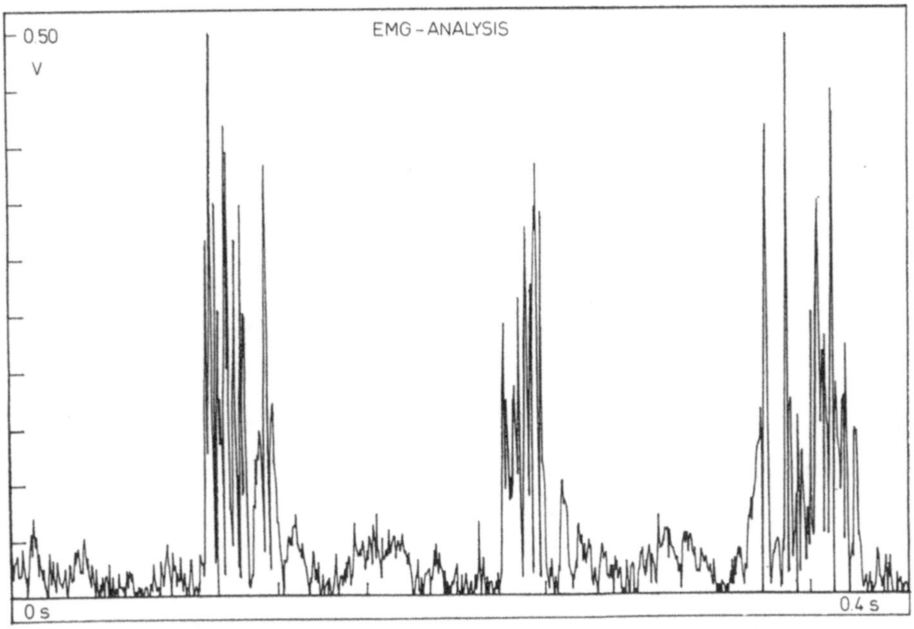

Fig. 3. As for Fig. 2, but at higher resolution

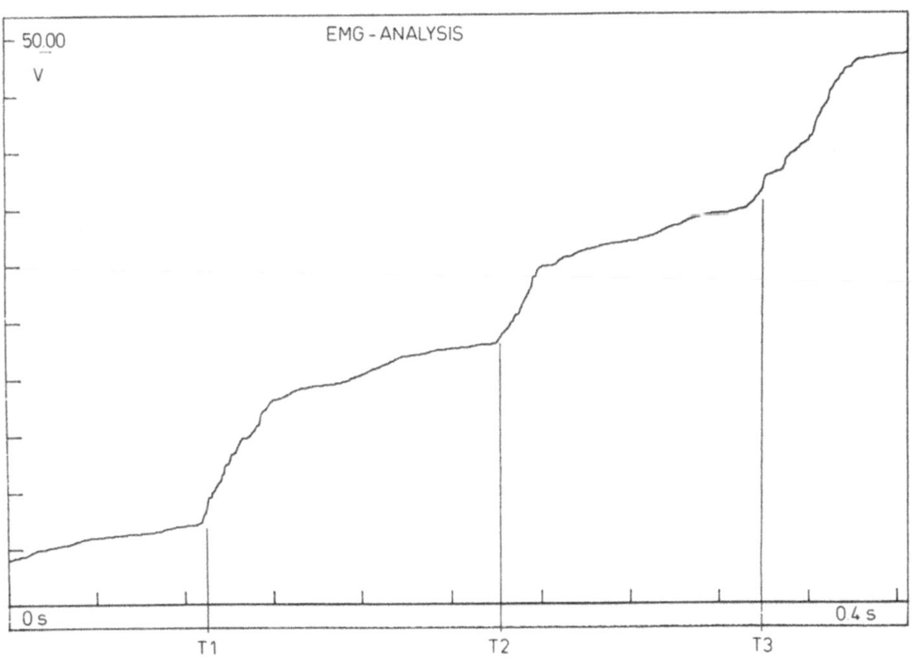

Fig. 4. Integration of signals from Fig. 3 (for details see text)

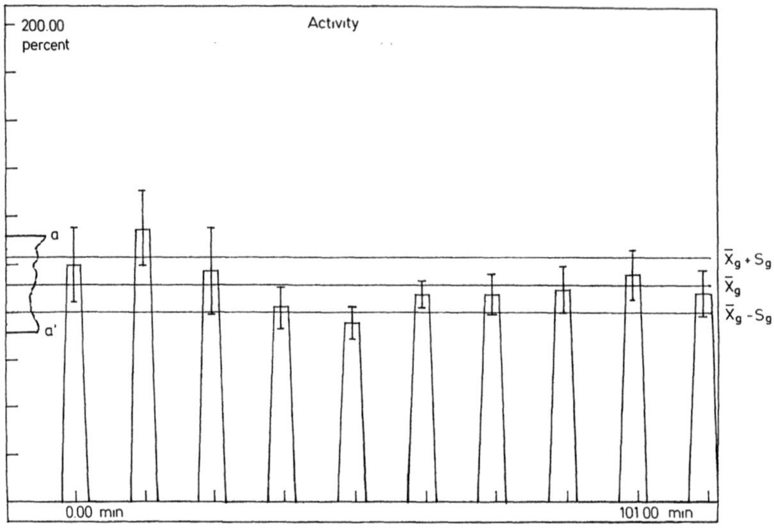

Fig. 5. Continuous tremor recording for 100 min with evaluation at 10-min intervals (*bars*). *Horizontal lines* indicate weighted means ± standard deviation; *a* and *a'* indicate limits of confidence intervals

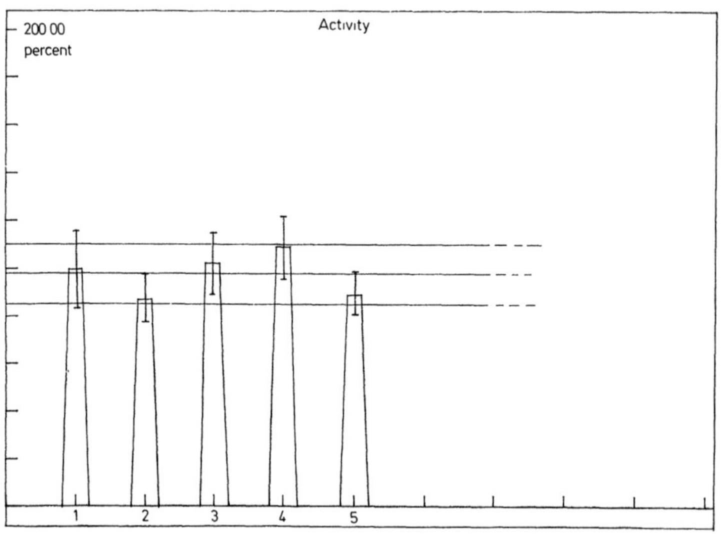

Fig. 6. EMG of a patient for 5 consecutive days with unchanged therapy (abscissa indicates EMG leads 1-5)

In Fig. 5 the result of a prolonged recording, lasting 100 min, is demonstrated. The computer analyzed EMG segments every 10 min. The columms represent the mean values of the analysis during this interval. The weighted mean value of the ten columns was 91% with a weighted standard deviation of ± 11.51%.

Figure 6 shows the measurements taken from the same patient on 5 consecutive days. In this case, the mean was 98%, with a standard deviation of ± 13%. Assuming a normal distribution, the confidence interval was established in the range of 100% ± 20%. We assume an increase or decrease in activity if the mean values of EMG recordings go beyond this interval in either direction. This simple and automatic tremor analysis appears to be highly suitable for the clinical survey of pharmacologic effects.

Technique

We used concentric needle electrodes. In comparison, the results obtained with surface leads appeared to be more reproducible when monitoring the course of the disease. However, we do not prefer this method for the following drawbacks:

1. Amplitudes are lower with skin electrodes than with needle electrodes. This changes the signal to noise ratio and affects muscular activity constants negatively.
2. The transition to zero tremor is ill defined with skin electrodes, so that it is difficult to determine group duration of individual tremor activities.
3. Large area leads, as in the case of finger extensors, may lead to superimposition of antagonist activity.

All of these reasons argue in favor of needle electrodes for monitoring activity. In the electromyogram, tremor activity is manifest as group activity which is fairly constant in the case of parkinsonian tremor. It is separated from subsequent action potentials by noise with its characteristic small amplitudes. Activity was monitored in the same muscular region each time. The potentials from areas with the highest reproducible amplitudes were evaluated. The experimental setting was always identical. Patients were requested to maintain regular time intervals between drug ingestion and EMG monitoring. Whenever they were exposed to great physical or mental stress, at home or on their way to the clinic, monitoring was postponed. There was no clear evidence of on-off phenomena during any of the examinations.

Patients and Therapy

So far, we have studied 16 patients (Table 1).

Monotherapy

The group treated with budipine was given daily doses of between 3 X 10 and 3 X 20 mg, with gradual increases.

Table 1. Patients treated with budipine

Number of patients monitored	16
Once	7
Several times	9
Treatment of patients monitores several times	
Receiving monotherapy	5
Extrapyramidal tremor	4
Essential tremor	1
Combination therapy with Madopar (extrapyramidal tremor)	4

Combination Therapy

The drug partner in combination therapy was levodopa (Madopar). When adding budipine, individual levodopa doses where maintained. In one female patient with essential tremor, the daily dose of budipine was increased to 4 × 20 mg. Continuous EMG recordings were obtained in several patients. With respect to the criteria mentioned previously and unchanged therapy, we confirmed the constant tremor group activity (Figs. 5 and 6).

Several patients who had undergone initial controls could not be followed up for reasons that were unrelated to the study. Some did not reappear for control, others reported subjective side effects, ascribing them to the drug, whereupon therapy was discontinued.

Figure 7 shows the change in tremor as a result of treatment. The left-hand side gives the standard 100% initial situation (baseline). To the right of this are the results under monotherapy. This patient (patient 6, extrapyramidal tremor) at first appeared to respond well, but this was masked by severe side effects, prompting discontinuation of monotherapy. The essential tremor in patient 7 was not amenable to improvement.

Figure 8 ist an example of the changes in patients who had received budipine in addition to Madopar. There is a significant decline in tremor activity in all patients. The effect is also demonstrable clinically. Two patients are still taking the drug without any untoward effect (patients 1 and 5, dose mostly between 3 × 10 and 3 × 20 mg). Figures 8 (patient 1) and 9 (patient 5) show the positive influence of prolonged combination therapy. Figure 9 (patient 5) confirms this result, with tracings in two muscles. This patient has now been given budipine for 15 months. There is no more evidence of tremor clinically, nor is it palpable, so that we discontinued doing further EMGs.

Side Effects

Patient 13 complained regularly of dry mouth.

Patient 7 complained of lack of appetite and gastric disorders; therapy was discontinued after 4 months.

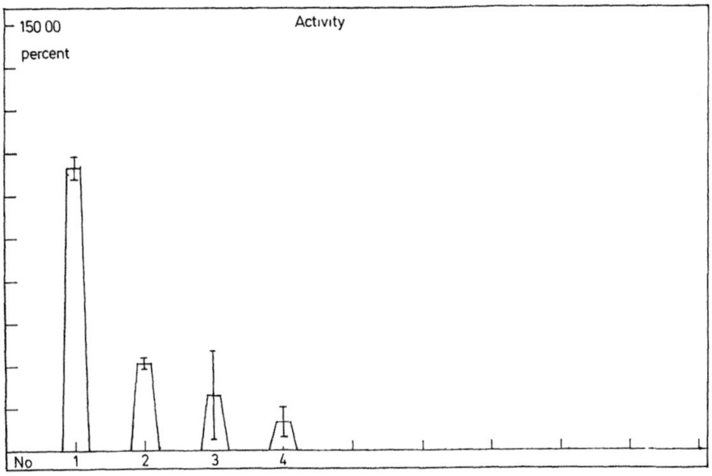

Madopar	500	375	375	375	mg/day
Budipine	-	30	30	30	
PK-Merz	-	-	-	200	

Fig. 7. Decline in tremor activity under budipine monotherapy (patient 6, three EMG leads in right m. extensor carpi ulnaris)

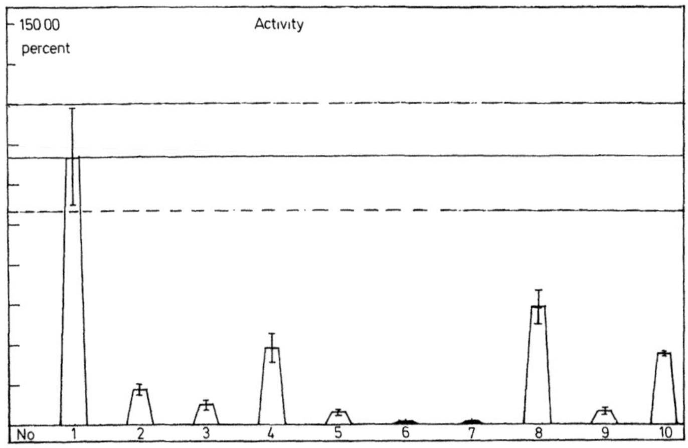

Madopar	500	500	500	250	500	500	500	500	500	500	mg/day
Budipine	-	40	50	80	100	60	60	30	30	30	
Rivotril	-	-	-	-	-	-	-	1 5	1 5	1 5	

Fig. 8. Decline in tremor under combined budipine and Madopar therapy over 1 year (patient 1, ten EMG leads in right m. rectus femoris)

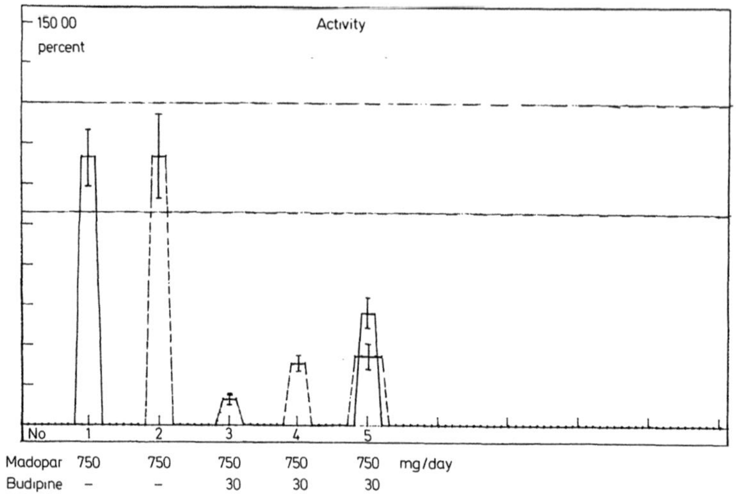

Fig. 9. Decline in tremor in two muscles under combined budipine and Madopar therapy (patient 5, six EMG leads: *full line* left m. extensor digitorum communis; *broken line* left m. tibialis anterior)

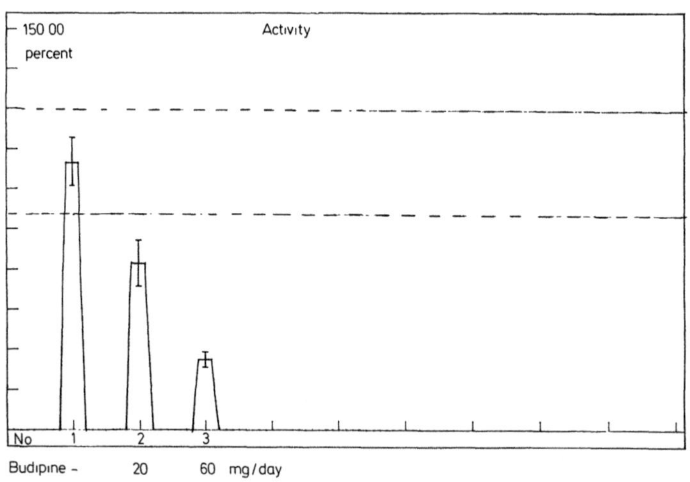

Fig. 10. Rigidity of extended muscle under combination therapy (patient 14, four EMG leads in left m. biceps brachii)

Patient 6 therapy was discontinued because of sleep disturbances, gastric disorders, and psychomotor restlessness. This patient had been taking a variety of other drugs. On withdrawal of budipine, only the stomach condition improved.

Patient 1 reported cramps in his right leg (probably unrelated to budipine, responding well to Rivotril). There was a marked psychotropic effect under budipine. Antidepressive treatment which had been necessary for many years was no longer required. This patient has now been taking budipine for 1.5 years with very good success.

Rigidity

Monitoring of rigidity with concentric needle electrodes is problematic. The results obtained in the resting and extended muscle are ambiguous, the reason being that voluntary muscular activity is masked by conscious or subconscious countertension. So far, we have failed to differentiate between the two. Hence, fluctuations in activity of more than 50% are possible with rigidity. Figure 10 shows the vague influence of budipine on rigidity. Clinically, rigidity appears to be improved, and this impression is confirmed by the EMG tracings, but, statistically, the result is not significant.

Conclusions

The computerized EMG data of tremor analysis are in good agreement with the clinical picture (Table 2). The muscle potentials derived by observation of the criteria described here permit one to draw an exact picture of tremor activity under treatment. Extrapyramidal tremor responded well to combination treatment with Madopar and budipine, but so far we have failed to demonstrate an effect of budipine monotherapy. The latter also applies to essential tremor.

Table 2. Results of budipine therapy

	Patients	Improved	Unchanged	Dropouts
Combination therapy with Madopar				
Extrapyramidal tremor	4	3	3	1
Monotherapy with budipine				
Extrapyramidal tremor	4	1	3	4[a]
Essential tremor	1		1	1

[a] Two patients withdrew because of side effects; two patients died during the study.

Vigilance Study Using Electrooculographic and Electroencephalographic Methods After Acute Budipine Administration in Healthy Subjects

K. Schaffler

Introduction

Administration of antiparkinsonian drugs in humans is often succeeded by a complex pattern of mental, motor, and vegetative effects. To elucidate these interactions, reliable and objective multidimensional approaches to different physiologic subsystems are required (Meyer-Delius and Schaffler 1975; Schaffler et al. 1975). An outstanding aspect of ambulant treatment with such drugs is possible impairment of vigilance, motor systems, and performance, because of their importance for everyday life, working and car driving.

The need for a simultaneous check of different psychophysiologic functions related to vigilance led to the design of the computerized oculodynamic test (ODT) (s. Fig. 1). The ODT (Schaffler 1982; Schaffler et al. 1981, 1982, to be published) simultaneously monitors the oculomotor system, operational parameters (such as complex choice reaction time, number of correct answers) as well as data from the cardiorespiratory system. For drug studies, the on-line monitoring of saccadic eye movements seems to be one of the most objective and effective methods, owing to the independence of this type of eye movement from exercise and motivation (Kornhuber 1978; Mackensen 1958; Vossius 1960/1961; Westheimer 1954).

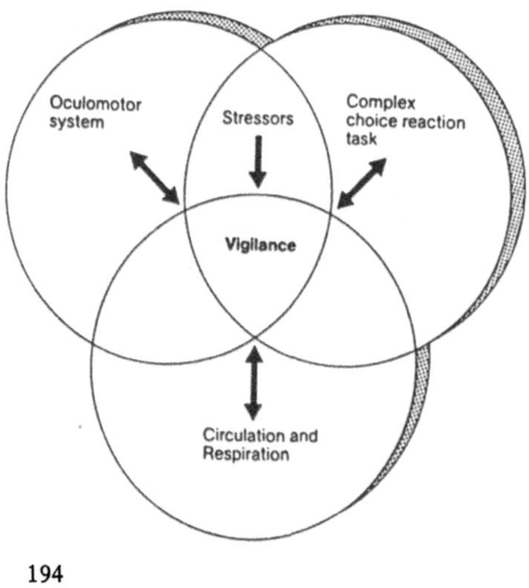

Fig. 1. Multidimensional approach to changes of vigilance

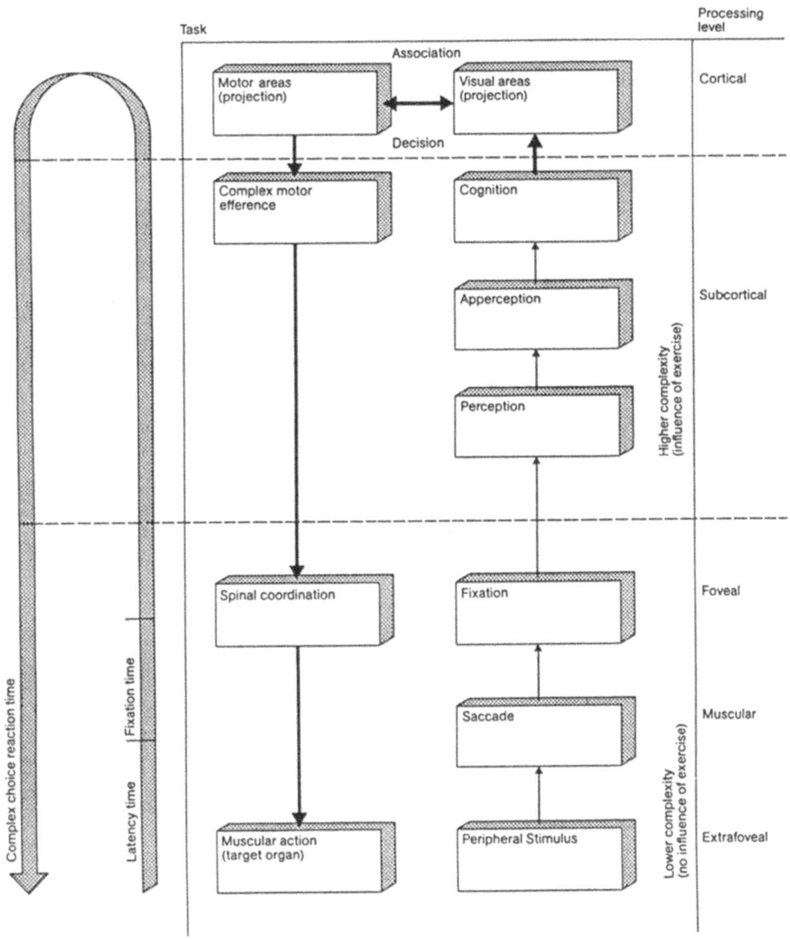

Fig. 2. Information processing of the oculomotor system

Saccades are rarely used for drug research, nevertheless the oculomotor system offers a lot of advantages for pharmacologic studies, such as:

Access to essential reflex mechanisms

Access to complex ascending and descending information processing of the CNS

Access to essential preprogrammes processes of the CNS

Access to a part of the brain (the eye) by "simple" experimental methods

Access to basic motor functions

Access to retinal functions (Fig. 2)

To get information on drug influence on "nonspecific" vigilance, computerized EEG was simultaneously run with the more specific and performance-related ODT.

Experimental Design and Methods

The budipine study with the ODT and the computerized EEG was performed in nine healthy, male volunteers (mean age 25.2 years; mean body weight 71.3 kg). It was a randomized, placebo-controlled, three-way, single-blind, crossover trial design. Treatment consists of an acute application of placebo, 30, and 50 mg budipine. The acute drug administration phases were separated by a 4-day washout period. Each subject was assessed three times on a pretest day for ODT training and laboratory adaption – to reach an optimal performance level for the operational data (complex choice reaction parameters, not necessary for the oculomotor parameters with reflex character). On each assessment day, the subjects ran a prevalue test and three more tests 60, 150, and 360 min after drug administration at the same time of day, to take account of circadian characteristics of the relevant physiologic parameters.

Oculodynamic Test

All subjects were asked to perform the computer-assisted ODT (Fig. 3). This test monitors the oculomotor system and parts of the general information processing of the CNS, with its global perceptive, apperceptive and cognitive components. Simultaneously, different on-line measurements of circulation and respiration as well as of operational

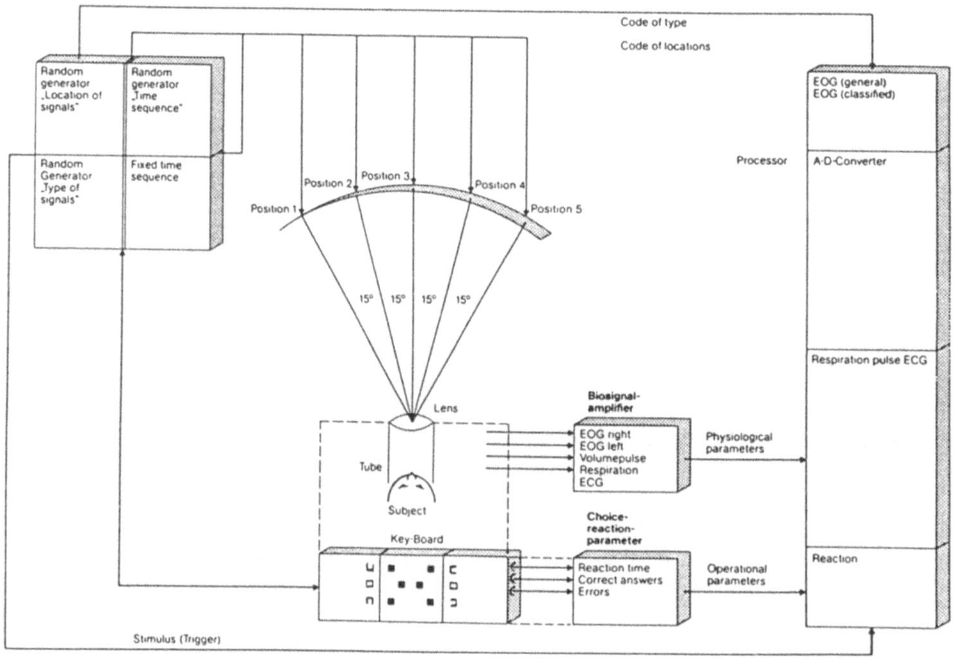

Fig. 3. Principles of the oculodynamic test

parameters were run. Details on the method are described elsewhere (Schaffler 1982; Schaffler et al. 1981, 1982).

A single ODT schedule was as follows:

5-min rest period

15-min working period (during which about 690 signals are presented)

2-min rest period.

The following data from the subjects were processed by a minicomputer (Varian V 73):

A. The electrooculogram of the left and right eye (Fig. 4)

1. The mean latency time of the left and right eye (i.e., the time between the appearance of the light signal in the periphery of the retina and the beginning of the eye movement for fixation of the target)

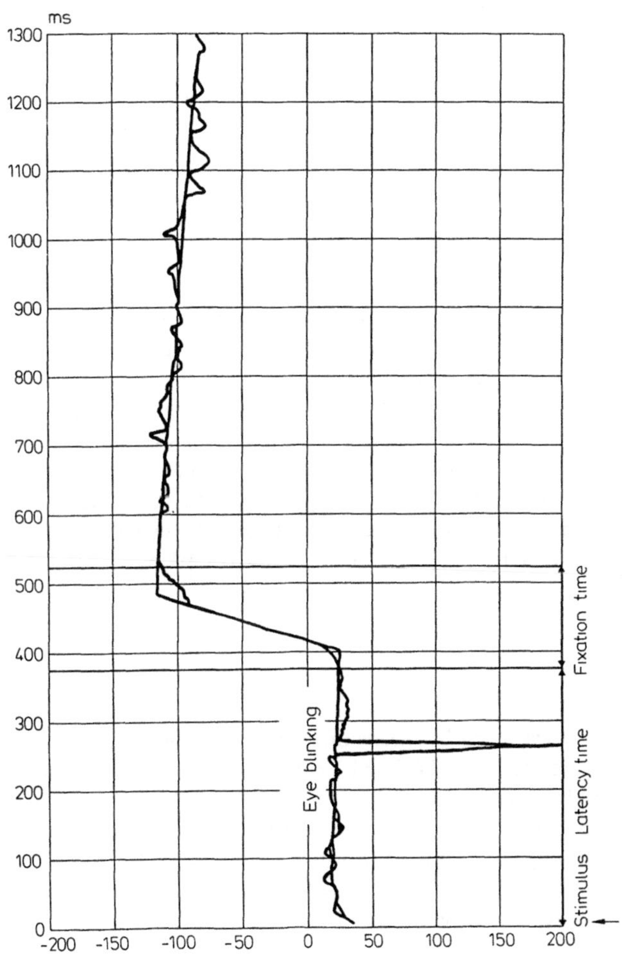

Fig. 4. Electrooculogram showing a simple movement of the left eye, 45°

2. The mean angular velocity of the right and left eyes saccadic movements to the signals
3. The accuracy of the fixation amplitude that could be held over one test procedure (related to the first 5 min of the working period)

B. Operational data (Fig. 3)
1. The mean reaction time in ms (responding to the different light signals in type on a keyboard)
2. The mean number of correctly idendified signals

C. Simple cardiovascular parameters (photoplethysmographic monitoring by transmission of the ear)
1. The mean pulse rate (beats/min)
2. The mean pulse arrhythmia, as a measure of the mean variability of the single pulse sequences (ms), standardized to the mean pulse frequency (beats/min)

D. Simple respiratory parameters with the aid of a nasal thermistor sensor
1. The mean respiratory rate (respirations/min)
2. The arrhythmia of the respiration

Computerized EEG

After each ODT session (0, 60, 150, 360 min) an on-line computer spectral analysis of the EEG was run for "eyes open" and "eyes closed". Bipolar EEG leads from O_2/C_Z were taken from the subjects in a supine position and fed into the computer after amplification (time constant 0.3, filter 30 Hz). After analogue-digital conversion (150 Hz) a fast Fourier transform (FFT) was made by segmentation of the EEG signal into 5-s intervals and linear averaging over a period of 20 s. Relative power and dominant frequency were computed in the following four frequency bands: delta 0.5-3.5 Hz; theta 3.5-7.5 Hz, alpha 7.5-12.5 Hz; and beta 12.5-30 Hz. Artifacts like eye-blinking and from electromyogram (EMG) are rejected automatically by the computer.

Results

The findings were statistically evaluated by means of appropriate descriptive and analytical methods (computer programs DESTAT and VARAN), based on the following scheme of a three-way analysis of variance:
1. Subjects/treatments/measuring times with repeated measurements (ODT)
2. Subjects/treatments/measuring times without repeated measurements (EEG)

If the null hypothesis was rejected, the findings were tested for significance by means of Scheffe's sequential test (Scheffe 1959).

Parameters of the ODT

Electrooculography. Table 1 gives the total values of the latency period of the saccadic eye movements, as means for all angles (x = 15°, 30°, 45°, and 60°) of the horizontal axis, in response to the peripheral retinal stimuli (i.e., the interval between appearance of the visual symbols and the beginning of the saccadic movements). The period tended to be a little shorter after both budipine dosages related to the placebo baseline (nonsignificant).

The angular velocity of the saccades, expressed as mean for all angular positions (i.e., the rate at which the eye moved to the target), could not be statistically differentiated from placebo in either budipine application. If the saccadic eye movements are classified (Table 2) in horizontal angles of view (15°, 30°, 45°, and 60°), budipine administration demonstrated a highly significant shortening of latency (1% level) for the two maximal positions of the saccades (45° and 60°) versus placebo.

Operational Parameters. The data from the integrated complex choice reaction part of the ODT system (Table 3) supported the view of indifferent behaviour of budipine (placebo-like). For the complex choice reaction time, no difference could be seen versus placebo, whereas the number of correct answers showed a marked, but nonsignificant trend toward improvement under budipine administration.

Cardiorespiratory Data. These simple but important, vegetative parameters were not affected in acute budipine treatment, related to placebo baseline.

Computerized EEG Data

Neither the results of the relative power nor of the dominant frequency in the EEG (Table 4) showed any outstanding trends different from placebo baseline for "eyes closed" in both budipine applications. In contrast for "eyes open" the 50-mg budipine preparation revealed a small, but significant augmentation of the relative power of the delta band, and a significant reduction of the beta band compared with placebo. The dominant frequency demonstrated and indifferent pattern of all medications.

Table 1. ODT results of electrooculographic parameters (unclassified)

Measurement	Medication	Mean value	Standard deviation	Significance (medication) three-factorial
Latency time (ms)	Placebo	380.0	20.0	N.S.
	Budipine 30 mg	376.7	17.0	
	Budipine 50 mg	376.9	18.5	
Angular velocity (°/s)	Placebo	470.7	89.1	N.S.
	Budipine 30 mg	471.4	89.0	
	Budipine 50 mg	469.1	94.4	

N.S., not significant.

Table 2. ODT results of electrooculographic parameters (classified)

Horizontal angles of view	Medication	Latency time (ms)		Significance (medication) three-factorial
		Mean value	Standard deviation	
	Placebo	374.9	16.2	N.S.
15°	Budipine 30 mg	374.6	15.2	
	Budipine 50 mg	375.0	14.0	
	Placebo	379.4	16.2	N.S.
30°	Budipine 30 mg	376.8	14.5	
	Budipine 50 mg	376.1	14.9	
	Placebo	384.2	19.4	1/3*
45°	Budipine 30 mg	382.0	17.5	
	Budipine 50 mg	377.8	20.8	
	Placebo	391.5	26.2	1/2*
60°	Budipine 30 mg	383.1	24.6	
	Budipine 50 mg	383.9	26.9	

N.S., not significant; * $P \leqslant 0.01$.

Discussion

Oculodynamic Parameters

The latency period, which should essentially be regarded as a polysynaptic retinomuscular reflex period (Kornhuber 1978; Mackensen 1958; Vossius 1960/1961; Westheimer 1954) of the electrooculogram, demonstrated no significant differences from placebo baseline after acute budipine administration (for both the 30-mg and 50-mg preparation). This fact led to the conclusion that the drug in the acute application regimen should not have a negative effect on polysynaptic reflex time (neuronal conduction and transmission time) in healthy volunteers.

Table 3. ODT results of operational parameters

Measurement	Medication	Mean value	Standard deviation	Significance (medication) three-factorial
Total reaction time (ms)	Placebo	854.6	65.1	N.S.
	Budipine 30 mg	851.8	63.7	
	Budipine 50 mg	860.7	67.1	
Correct answers (min^{-1})	Placebo	27.6	10.3	N.S.
	Budipine 30 mg	29.1	11.8	
	Budipine 50 mg	29.0	10.2	

N.S., not significant.

Table 4. Results of computerized EEG Spectral analysis (eyes open)

Measurement	Medication	Mean value	Standard deviation	Significance (medication) three-factorial
Relative power				
Delta (%)	Placebo	28.4	4.6	1/3*
	Budipine 30 mg	29.3	5.6	
	Budipine 50 mg	32.2	5.6	
Theta (%)	Placebo	21.3	3.3	N.S.
	Budipine 30 mg	21.0	3.5	
	Budipine 50 mg	20.9	3.7	
Alpha (%)	Placebo	22.8	5.5	N.S.
	Budipine 30 mg	22.8	5.7	
	Budipine 50 mg	21.7	7.3	
Beta (%)	Placebo	27.5	4.4	1/3*
	Budipine 30 mg	26.8	6.7	
	Budipine 50 mg	25.2	5.2	
Dominant frequency				
Delta (Hz)	Placebo	1.3	0.6	N.S.
	Budipine 30 mg	1.1	0.5	
	Budipine 50 mg)	1.2	0.6	
Theta (Hz)	Placebo	5.1	1.3	N.S.
	Budipine 30 mg	5.0	1.2	
	Budipine 50 mg	4.9	1.0	
Alpha (Hz)	Placebo	9.5	1.1	N.S.
	Budipine 30 mg	9.2	1.2	
	Budipine 50 mg	9,5	1.3	
Beta (Hz)	Placebo	13.9	1.6	N.S.
	Budipine 30 mg	13.8	1.5	
	Budipine 50 mg	14.2	2.7	

N.S., not significant; * $P \leqslant 0.05$.

An adequate explanation could be given for the angular velocity of the saccades. Effects on this parameter are largely governed by the function and innervation of the musculi recti of the eye. The eye muscles have the highest innervation density of all striated muscles of the body (smallest motor unit approximately 1:7 compared with approximately 1:1200 in the musculi gastrocnemii), as well as a small intrinsic mass with relatively few additional viscoelastic effects. Therefore, eye muscles may be regarded as very sensitive indicators of any effects on neuromuscular transmission or muscular function. As a result of this study, budipine revealed no effect on sensitive striated muscle function in either acute dosage. Identical features concerning neutral drug effect could be derived from the complex choice reaction parameters (operational data) and cardiorespiratory data after single dose application of budipine in volunteers.

Computerized EEG Data

For the acute administration regimen, no typical drug pattern could be demonstrated in the computer spectral analysis of the EEG (eyes open). A slight trend to a classical antidepressive type of pharmaco-EEG could be detected for the 50-mg preparation of budipine, but no dramatic effects on "unspecific" vigilance could be monitored for the acute administration in healthy subjects in relation to placebo application.

Conclusions

In an acute single-blind, crossover study with nine healthy subjects, neither 30-mg nor 50-mg acute doses of budipine revealed any negative effects in relation to placebo baseline values on unspecific vigilance (computerized EEG) or on specific performance data (car-driving or working), monitored by a very sensitive psychophysiologic test, the ODT, which gives a complete set of information on the oculomotor and complex choice reaction systems, as well as on simple cardiorespiratory parameters. These results, derived from healthy subjects, could be interesting for clinical as well as for ambulant treatment of patients with parkinson's disease, espesially because of the lack of impairment of important psychophysiologic and motor parameters.

References

Kornhuber HH (1978) Blickmotorik. In: Gauer OH, Kramer K, Jung R (eds) Physiologie des Menschen, vol 6. Urban and Schwarzenberg Munich, pp 380-381, 391-392

Mackensen G (1958) Die Geschwindigkeit horizontaler Blickbewegungen. Graefes Arch Ophthalmol 47: 160

Meyer-Delius H, Schaffler K (1975) Stress and Strain bei visueller Perzeption. Z Arb Wiss 1: 33

Schaffler K (1982) The action of Clobazam and Diazepam on computer-assisted tests of muscle activity: Oculomotor effects. Drug Dev Res S1: 39-46

Schaffler K, Lincke P, Meyer-Delius H (1975) Der Einfluß von Aufmerksamkeitsanforderungen auf Augeneinstellbewegungen, visuelle Perzeption und Rhythmik der Kreislaufreaktion. Ärztl Prax 70: 2679

Schaffler K, Hörmann E, Arnold H (1981) A pilot-study to evaluate the effect of acute and long-term administration of Cyclandelate on the vigilance of subjects submitted to hypoxic conditions. Drug Res 31/1, 6: 1032-1035

Schaffler K, Arnold H, Hörmann E (1982) Vigilanzverhalten bei Interaktion des serotonin-selektiven Antidepressivums Zimelidin mit parenteral verabreichten Alkohol. Drug Res 32/2, 8: 845-852

Schaffler K, Hörmann E, Arnold H (to be published) Dose-effectiveness-study on vigilance processing data from the oculomotoric, cardiorespiratory and choice-reaction system after acute administration of midazolam – a short-acting benzo-diazepine. Br J Clin Pharmacol

Scheffe H (1959) The analysis of variance. Wiley, New York

Vossius G (1960/1961) Das System der Augenbewegungen. Z Biol 27: 112

Westheimer G (1954) Eye movement responses to a horizontally moving visual stimulus. Arch Ophthalmol 52: 932

Budipine in the Treatment of Extrapyramidal Disorders Not Related to Parkinson's Disease

D. Müller

Introduction

Budipine is a new drug in the treatment of Parkinson's disease. According to preliminary experience, it is effective against tremor and rigidity as well as against akinesia. It was demonstrated that this new diphenylpiperidine derivative is effective in both monoaminergic and cholinergic neurotransmitter systems. It seems to be able to stimulate dopaminergic, noradrenergic, and serotonergic activity and to act as an anticholinergic drug (Menge and Brand 1982). Thus, it is able to influence all symptoms of Parkinson's disease. In our present study, the results of budipine therapy in patients suffering from other extrapyramidal disorders, such as spasmodic torticollis dystonicus, choreoathetoid hyperkinesia, intention hyperkinesia, and essential tremor, are reported.

Patients

Our study contained seven patients with spasmodic torticollis, three with choreoathetoid hyperkinesia, three with intention hyperkinesia, and four with essential tremor.

Spasmodic Torticollis

The group of patients suffering from spasmodic torticollis included four men aged 23, 38, 40, and 51 years. The duration of illness was 3.5-6 years before starting treatment. All the patients had been treated unsuccessfully with other drugs several years before therapy with budipine started. Two patients had previously undergone stereotactic brain operations, which improved the situation, but did not free them of symptoms; both these patients still suffered from residual symptoms requiring treatment.

Aside from a subdepressive emotional state, no neurologic or psychiatric traits were noticeable in three patients. One patient suffered from mild postoperative unilateral symptoms with slight spasticity. Medical diagnosis found one patient with chronic bronchitis; the other three were found to be healthy. In addition, three further patients were treated with budipine temporarily.

Four cases involved long-term treatment: the shortest treatment was 1 year 10 months; the longest 3 years 5 months. The initial dose was 2 × 10 mg/day in three cases and 3 × 10 mg/day in one case. The dose of 2 × 10 mg/day was increased to 3 × 20 mg/

day maintained at this level 1 year 6 months. It has since been reduced to 3 × 10 mg/day. The maintenance dose for two patients is 2 × 20 mg/day; in one particularly favorable case, the dose is presently 2 × 5 mg/day.

Three of four patients were treated with 2 × 5 to 3 × 10 mg/day Dartal (a thiopropazate) in addition. The fourth patient was treated in addition with 3 × 25 mg/day baclofen and 3 × 50 mg phenytoin, daily because of side effects caused by an operation.

Choreoathetoid Hyperkinesia

Very different reasons for damage existed in three of the patients with choreoathetoid hyperkinesia whom we treated with budipine; the youngest patient (23 years old at start of treatment) had suffered a severe craniocerebral trauma with hemiparesis on the right side 2 years previously. After regression of the paresis, there occurred choreiform hyperkinesias, primarily in the arm, in conjunction with constant resting spastic paresis. A stereotactic operation had reduced part of the hyperkinetic symptoms; however the residual symptoms required further treatment.

In the second patient (a 43-year-old man whose sickness began at 33 years), toxic brain damage is assumed to be the cause. In the third patient (a 62-year-old man), severe encephalitis was recorded early in childhood. This patient had received relief from hyperkinesia, primarily in the arm, by means of a stereotactic brain operation before starting budipine treatment. The residual symptoms, which had intensified in recent years, necessitated further drug treatment. All three patients had been unsuccessfully treated for years with dopa preparations, other anticholinergics, GABA, dantrolene sodium, and psychotropics. These patients have also undergone long-term treatment for 1.5-3 years. Therapy was begun with low doses, as was the case with those patients suffering from torticollis; however, higher maintenance doses of 3 × 20 mg/day budipine have become necessary. There was no other supportive medication.

Intention Hyperkinesia

There were three patients with intention hyperkinesia. Patient 1 started treatment at the age of 27 years. Because of congenital spinocerebellar degeneration, he has suffered since childhood from severe ataxia, primarily in the leg and from severe tremor, primarily in the arm, as well as in the trunk and head.

A unilateral stereotactic thalamotomy 3 years before starting budipine treatment decreased the symptoms on the right side. The residual symptoms especially those on the unoperated left side, have been resistant to every drug combination conceivable. This patient has been received 4 × 10 mg/day budipine since March 1980, which he tolerated without any problems. Simultaneously, he is receiving 3 × 800 mg piracetam (Nootrop) per day.

In a parallel case, the patient (about the same age) has been undergoing treatment for 4 weeks at the time of writing. This patient, too, suffered from congenital intention hyperkinesia; however, the symptoms were less severe and occurred exclusively in the arms and trunk. He has also undergone every conceivable drug combination without success over a period of several years. One week before starting budipine treatment, the patient underwent a stereotactic brain operation on the right side. The more severe symptoms on the patient's left side were thus caused to regress. After subsidence of the immediate postoperative symptoms, especially of cerebral edema, budipine treatment was started with intravenous infusions of 30 mg/day. The dose was increased to 60 mg/day on the day 3; 10 days after starting budipine treatment, the drug was administered orally at 3 X 20 mg/day.

The third patient ist a 47-year-old woman with disseminated encephalomyelitis who also suffered from cerebro-organic seizures. When budipine treatment was begun, the patient was suffering from paraparesis of both legs, severe ataxia, primarily in the arms and trunk with intention tremor, and urinary incontinence with chronic urinary tract infection. A stereotactic thalamotomy on the left side of the brain caused the tremor symptoms to regress on the right side of the body; however, the left side of the body could not be influenced. In view of the severe intention tremor, there was a need for further treatment. Since December 1978, the patient has been receiving budipine treatment; this was initially administered carefully at small doses of budipine and then increased to a maximum dose of 3 X 40 mg/day.

Essential Tremor

Two patients aged 56 and 66 years with therapy-resistant essential head tremor were temporarily treated with budipine. Both patients were women who had suffered from this head tremor since their youth. Both had undergone long-term treatment with dopa preparations, anticholinergics, psychotropics, as well as acupuncture and other unconventional methods. In both cases, unilateral stereotactic thalamotomy had reduced tremor activity for only 6-8 months, so that further treatment became necessary, since both women refused a second operation. They received budipine at doses of up to 3 X 20 mg/day. No other drugs were administered during treatment with budipine.[1]

Results

The results of the study are summarized in Table 1.

Spasmodic Torticollis

Treatment with budipine has caused a marked decrease in symptoms of torticollis in the four patients mentioned. In addition, all the patients had received supportive medication with thiopropazate or baclofen and phenytoin. Previous treatment with these drugs over a long period in combination with other different anticholinergics did not produce any satisfactory results. The patients suffered particularly from autonomic and gastrointestinal side effects from these other anticholinergics. These side effects

1 In the meantime, two new male patients with generalized essential tremor have received budipine treatment.

Table 1. Results of budipine treatment in extrapyramidal disorders

Syndrome	Number of cases	Result[a] *A*	*B*	*C*	*D*	*E*	*F*
Spasmodic torticollis	7	3	1				3
Choreoathetoid hyperkinesia	3	2		1			
Intention hyperkinesia (degenerative)	2	1	1				
Intention hyperkinesia (multiple sclerosis)	1			1			
Essential tremor	4	2			2		

[a] *A* permanent good improvement; *B* moderate improvement; *C* transient improvement; *D* no change; *E* worsening; *F* ceased because of side effects

disappeared completely under budipine treatment. All four patients were able to work again. An attempt is being carried out in the meantime to lower the maintenance dose of budipine.

Budipine treatment was tried temporarily in three other patients, since they, too, had suffered markedly from the side effects of other anticholinergics; however, these patients also reported negative side effects (especially a subjective feeling of unrest) at 2-3 × 10 mg/day budipine; thus, treatment had to be stopped after a few days.

Choreoathetoid Hyperkinesia

In all three patients suffering from choreoathetoid hyperkinesia, the symptoms have regressed markedly under budipine treatment. One of the younger men became able to work in the meantime. There has been a marked improvement in the residual symptoms of the older patient; he no longer requires special care as he did before beginning treatment with budipine. In contrast to these two cases, the budipine effect decreased progressively about 2 years 6 months after beginning treatment in the 43-year-old man with toxic brain damage. In spite of increasing the dose to 4 × 30 mg/day, no regression in the symptoms occurred. After a total of 2 years 10 months, drug treatment was stopped and the patient underwent a stereotactic operation. Since the symptoms were still only slightly decreased, budipine treatment was started again 6 months later. Presently, the patient is receiving 4 × 10 mg/day budipine; he has responded well to this treatment for more than 8 months. None of the patients is suffering from any side effects caused by budipine. Budipine therapy is being continued in all three patients.

Intention Hyperkinesia

The "three" cases with intention hyperkinesia showed different responses to budipine treatment. One of the patients suffering from spinocerebellar degeneration revealed a surprisingly favourable improvement of intention tremor, beginning about 6-8 weeks after budipine treatment had started. He is attending a workshop for the handicapped where he works as a mechanic. The present maintenance dose is 3 X 10 mg/day budipine, which is tolerated without any side effects. The patient had been confined to a wheelchair up to August 1982. He is now able to walk a few steps unaided for the first time in 15 years.

The second patient showed no improvement initially. After increasing the dose to 3 X 20 mg/day with oral administration of the drug, a moderate regression of symptoms occured 3 months after starting budipine therapy. The dose was tolerated by the patient without any difficulties or side effects. The tendency toward improvement is continuing.

In the third patient, a 47-year-old woman with intention tremor and ataxia due to disseminated encephalomyelitis, only transient response to budipine was observed. In the first year of treatment there was initially marked alleviation of the intention tremor symptoms, primarily on the left side after the operation. As the demyelinating disease progressed, the effect on hyperkinesia became less and less. Budipine treatment was terminated after a few weeks in the spring of 1980 when the tremor symptoms on the left side became as severe as they had been earlier. No serious side effects had occurred from budipine, in spite of occasional high doses of 3 X 40 mg/day.

Essential Tremor

No influence on the tremor symptoms was achieved in either of the female patients with essential tremor during a 6-month treatment period. The medication with budipine was therefore terminated. No side effects were observed during treatment with budipine.[2]

Discussion

We have now had more than 3 years and 6 months experience with budipine treatment of extrapyramidal disorders not caused by Parkinson's disease. Budipine has proved to be very effective in patients with otherwise therapy-resistant extrapyramidal torticollis, therapy-resistant choreoathetoid hyperkinesia, as well as in two young men with congenital intention hyperkinesia.

2 In the meantime, two new male patients with severe essential tremor have been stabilized on budipine. Contrary to the failures reported in the study, the results are very favourable with both patients. One of them is able to resume work already.

Spasmodic Torticollis

With spasmodic or dystonic torticollis, the loss of cell function is assumed to occur in the striatum (Hassler 1953), in the anterior thalamic central nucleus (Hassler and Dieckmann 1970), and in certain pontine structures (vestibular nuclei, Deiters' nucleus, Saavedra 1979). It is further assumed that the pathophysiologic result thereof is a unilaterally pronounced hyperactivity of the pallidum and of pallidothalamic fiber bundles. The logical goal of drug therapy is thus to inhibit pallidofugal impulses; for this, anticholinergic and serotonergic agonistic drugs appear to be particularly suited. This also includes certain psychotropics. In comparison, treatment with dopa drugs, GABA, dantrolense sodium, and spasmolytic drugs has proved to be ineffective.

Compared with the effect of previously used anticholinergics on the hyperkinetic impulses in our patients, that of budipine proved to be at least as good: in three of four cases it was better; in one case it was substantially better. It is to be noted, however, that budipine was not administered as monotherapy to any of the four patients who were treated successfully. The successful treatment can definitely be traced to the supportive administration of thiopropazate. The combination with budipine, however, proved to be especially effective and to have only few side effects.

Choreoathetoid Hyperkinesia

The pathologic substrate in choreoathetoid hyperkinesia can be seen in a damaging of the striatum, as is similar with torsion dystonia and dystonic torticollis. Frequently, the external pallidum is affected in addition. Pathophysiologically, with this disease there is also disinhibition in pallidothalamic fiber systems and in motor centers of the thalamus (Müller 1983).

Intention Hyperkinesia

Intention hyperkinesia is caused by congenital, posttraumatic, and encephalitic (multiple sclerosis) lesions of the nucleus dentatus and of cerebellar fiber bundles with run by subthalamic pathways to the ventro-oral thalamic nuclei. It may be that the serotonergic activity of budipine improves the disintegrated function of thalamic motor nuclei which are disinhibited by cerebellar damage. The transient response to budipine treatment in patient 3 is obviously due to the progressing encephalitic disease (multiple sclerosis).

Essential Tremor

The diverging results of budipine therapy in essential tremor can not be explained satisfactorily and remain open to further discussion, since the etiology and pathophysiology of this disorder is still unknown. In 1972, Riechert and Richter reported loss of parvicellular neurons in the striatum.

Side Effects

Finally, budipine has the special advantage of causing fewer side effects than do comparable drugs, according to our experience. We were thus able to benefit from a greater dose margin. This opinion is supported by observations made of 18 parkinsonian patients treated with budipine. They were mainly patients who had suffered from side effects of other antiparkinsonian drugs. Among them is a 48-year-old man with a postencephalitic parkinsonian syndrome; he tolerated monotherapy of 4 X 30 mg/day budipine without any side effects for more than 18 months. Since then, this patient has tolerated a dose of 180 mg/day budipine, also without any side effects. According to my estimation, budipine is a valuable supplement to drugs previously used for treating extrapyramidal diseases.

Summary

The results of budipine treatment in seven patients with spasmodic torticollis, three with choreoathetoid hyperkinesia, three with intention hyperkinesia, and four with essential tremor are reported. Three of seven torticollis patients who were followed up in a long-term study from approximately 2 to 3.5 years showed excellent improvement, one patient moderate relief of symptoms. In three patients, treatment was discontinued because of side effects. Two patients with choreoathetoid hyperkinesia, two with intention hyperkinesia, and tow with essential tremor also showed good to moderate improvement. One patient, with choreoathetoid hyperkinesia and one with intention hyperkinesia caused by multiple sclerosis revealed only transient response to budipine. Two patients with essential tremor did not respond. In our experience, budipine caused fewer side effects than did other comparable drugs. In torticollis patients, budipine treatment was combined with administration of thioprozate which also showed fewer side effects than combination of thioprozate with other anticholinergic drugs. Thus, budipine is in our estimation a valuable supplement to the drugs previously used for treating extrapyramidal diseases.

References

Hassler R (1953) Extrapyramidal-motorische Syndrome und Erkrankungen. In: Jung R (ed) Neurologie, 4th edn. Springer, Berlin Heidelberg New York (Handbuch der Inneren Medizin, vol 5/3)

Hassler R, Dieckmann G (1970) Die stereotaktische Behandlung des Torticollis aufgrund tierexperimenteller Erfahrungen über die richtungsbestimmten Bewegungen. Nervenarzt 41: 743-487

Menge HG, Brand U (1982) Zusammenfassende Darstellung der Pharmakologie von Budipin, einem neuen 4,4-Diphenylpiperidin-Derivat für die Parkinson Therapie. Drug Res 32: 85-98

Müller D (1983) Stereotaktische Behandlung extrapyramidaler Krankheiten. In: Hopf HC, Poeck K, Schliack H (eds) Neurologie in Praxis und Klinik, vol 1. Thieme, Stuttgart, pp 6.82-6.87

Riechert T, Richter D (1972) Operative Behandlung des Tremors der Multiplen Sklerose und des essentiellen Tremors. Münch Med Wschr 114: 2025-2028

Saavedra JL (1979) Der Torticollis spasticus und seine stereotaktische Behandlung. Dissertation, University of Göttingen

Subject Index

Accelerometry 14
Acetylcholine 83
Acetylcholine esterase 83
ADL (activities of daily living) 44, 47
Adrenaline 7
AIM (abnormal involuntary movements) 60, 61, 73
Alpha$_2$-adrenoceptors 103
Akinesia 16, 40
Akinesia models 83
Alzheimer's disease 9
Amantadine 69, 84, 85, 130
Amphetamine 87
Amygdala 6
Anticholinergic drugs 2
Anticholinergics 83, 169
Apomorphine 23, 24, 87, 95, 108
Apomorphine ^{3}H 32, 61, 62, 63
Atropine 85
Automatic movements 44
Axial rotation 47

Ballistic movement 17
Basal ganglia 13, 17, 44, 50
Benserazide 79
Bentropine 160
Benzodiazepinergic 108
Bereitschaftspotential 18
Beta-hydroxylase 8
Beta-receptor blockers 142
Biperiden 84, 85, 176
Blink reflex 16
Blood viscosity 182
Bornaprin 179
Bromocriptine 23, 26, 31, 33, 64, 73, 75, 78, 79, 84, 85, 108

Carbidopa 79
CAT (choline acetyltransferase activity) 9
Catalepsy 83
Caudate 23
Caudate nucleus 6, 8
CCK-8 (cholecystokinin-8) 4, 10
Cerebral atrophy 54
Cerebral blood flow 182
Chloramphetamine 88
Choline acetyltransferase 7
Cholinergic systems 8
Cingular 6
Clonidine 87, 95, 103
CM 29–712 33
Complex choice reaction time 194
Complex visual reaction time 54
Computer tomography 52, 54
Corpus striatum 87
CQ 32–084 27, 33
CU 32–085 27, 33

Dantrolene sodium 208
DBH (dopamine beta-hydroxylase) 42
Dementia 4, 9, 32, 34
Deprenyl 64, 80
Depression 53
Diphenylpiperidine 82
Disseminated encephalomyelitis 205
Domperidone 68
DOPAC (dihydroxyphenylacetic acid 3,4-) 6, 108
Dopamine 4, 6, 13, 22, 79, 108
Dopamine ^{3}H 103
Dopamine receptors 22
Dopaminergic 108
Dopaminergic agonists 22, 24, 33, 68
Dopaminergic receptors 7, 22, 31, 109, 156
Dopaminergic systems 4, 6, 44
DOPS (dihydroxyphenylserine 3,4-) 79
Dorsal roots 14
D1 receptors 22
D2 receptors 23, 87
Drug-induced psychosis 55
Dyskinesia 22, 24, 26, 32, 34, 122, 153, 155, 158, 159, 169
Dystonic cramps 65

Early morning dystonia 26
ECT (electroconvulsive therapy) 2
EEG 52, 55, 196

EMG (electromyography) 14, 185
Electrooculogram 197
End-of-dose failure 75, 77, 153
Endorphin 7
Entorhinal cortex 6
Essential tremor 107, 147, 205
External pallidum 9

Fenfluramine 88
Flicker fusion therapy 178, 181
Fluctuations 32
Fluctuations in performance 32
Flupenthioxol 69
Freezing 41, 153
Frequency analysis 169
Frontal cortex 6, 9

GABA 71
GABAergic 108
GABAergic receptor 109
GABAergic systems 9
GAD (glutamic acid decarboxylase) 7, 9
Gamma system 15
Grünberger test 177

HA (homovanillic acid) 108
Haloperidol 23
Haloperidol ^{3}H 31, 83
HIAA (hydroxyindoleacetic acid, 5-) 108
Hippocampus 6, 8, 9
Histamine 8, 71
Histidine decarboxylase 4
HPLC (high pressure liquid chromatogram) 108, 113
HR, 5- 88
H-reflex 14–16
HVA 6
Hydroxydopamine 87
Hyperkinesia 204
Hypothalamic systems 6
Hypothalamus 4, 7, 26, 104

IA fibres 15
IA interneuron 16
Innominatocortical system 4
Intellectual disorders 54
Involuntary movements 58

Kinesie paradoxale 41

LA (locomotor activity) 93
Lergotrile 23, 27, 32
l-deprenyl 73, 75, 77
Levodopa 22, 31, 41, 44, 45, 50, 52, 58, 75, 79, 85
Levodopa long-term treatment 33
Levodopa therapy 53
Levodopa treatment 69
Lewy bodies 8
Limbic areas 4, 6
Lipotropin 7
Lisuride 23, 27, 33, 66, 69, 79
Locus coerulus 8, 79, 104
LON-954 83
Long latency reflexes 16
Long loop reflexes 13
Long-term course 53
Long-term effectiveness 58
Long-term therapy 52
Lost of levodopa response 34
L-threo-DOPS (L-threo-3,4-dihydrocy-phenylserine) 42
L-tryptophan 79

MAO (monoamine oxidase inhibitors) 80
Mass spectrometric 114
Medulla oblongata 8
Melanine 6
Mental disorders 52
Mesocorticolimbic system 6
Methionine-enkephalin 4, 7, 9
Microneurography 15
Migraine 107
Motor behaviour 44
Motor programme 14
Muscarinic receptors 9, 82
Muscle spindles 14

Neuroleptic drugs 31
Neuroleptic receptor 109
Neuroleptics 169
Nigrostriatal pathway 4, 7, 24, 44
Nigrostriatal system 6, 68
Nondopaminergic system 8
N-propylnoraporphine 26
Noradrenaline 7, 8, 71, 79, 87
Noradrenaline ^{3}H 103
Noradrenergic system 8
Norepinephrine 42
Nucleus accumbens 6, 7, 31, 104
Nucleus caudatus 108
Nucleus dentatus 208
Nucleus paranigralis 6

Oculodynamic test 194
Off phase 61, 62
On-off phenomena 34, 36, 73, 152, 155, 158, 159
On-off reactions 22, 58, 60
On-off responses 59

Orphenadrine 69
Oxotremorine 82
Opiate 108
Opiate receptors 109

Pallidum 23, 208
Peak dose dyskinesia 77
Pentoxifylline 69
Peptidergic systems 9
Pergolide 27, 33, 64, 79
Peroneal nerve 16
Perphenazine 83
Pharmacotoxic psychosis 55
Phosphodiesterase 69
Physiologic tremor 166
Physiotherapy 44, 45
Physostigmine 63, 82
Piracetum 204
Piribedil 68, 104
Platelet aggregation 182
Porsol swim test 93
Postural reflexes 45, 47
Postural tremor 14
Prefrontal cortex 6
Procyclidine 179
Prodipine 80, 107, 142
Prolactin 26, 32
Putamen 9, 23
Psychotic episodes 34

Quipazine 88

Raphe 8
Rauwolfia serpentina 2
Readiness potential 18
Reserpine 83, 109
Rigidity 15, 40, 193
Rigidity models 83

Schizophrenia 169, 176
Septum 6
Serotonergic 108
Serotonergic receptors 34, 88, 109
Serotonergic systems 8
Serotonin 79, 83, 108, 156
Serotonin ^3H 103
Shortening reaction 15
Sinemet 64
Spasmodic torticollis 203
Spasticity 13
Spinal cord 13
Spinocerebellar degeneration 204, 207
Spiperone ^3H 7, 31
Spiroperidol 23, 87
Stereotactic operations 204
Stereotactic subthalamotomy 167, 168
Stereotactic surgery 41, 44
Striatum 4, 6, 7, 24, 26–28, 31, 32, 104
Substance P 4, 10
Substantia innominata 9
Substantia nigra 6, 10, 23, 31, 104
Sural nerve 16
Synaptosomes 87

Tardive dyskinesia 176
Tetrabenazine 83
Thalamotomy 40, 204, 205
Thalamus 208
Tiapride 69, 73, 78
Tonic vibration reflex 15
Torsion dystonia 208
Trancylcypromine 80
Trihexyphenidyl 69, 160
Tremor 14, 68, 140, 142, 159, 163, 185, 190
Tremor analysis 189
Tremor measurements 163
Tremor models 82
TRH 4, 7, 8, 31
Tyramine 87

Urinary retention 141, 149

Ventrotegmental area 6, 10, 104

Wearing-off 22, 26, 73
Webster rating scale 123, 130, 140, 144, 153, 155, 158, 169, 172

Yohymbine 87, 105
"Yo-yo"-symptoms 75, 78

GPSR Compliance

The European Union's (EU) General Product Safety Regulation (GPSR) is a set of rules that requires consumer products to be safe and our obligations to ensure this.

If you have any concerns about our products, you can contact us on ProductSafety@springernature.com

In case Publisher is established outside the EU, the EU authorized representative is:

Springer Nature Customer Service Center GmbH
Europaplatz 3
69115 Heidelberg, Germany

Batch number: 10684222

Printed by Printforce, the Netherlands